GENERATIVE ENERGY

RESTORING THE WHOLENESS OF LIFE

THIRD EDITION

Raymond Peat, PhD

Ray Peat® Publishing
www.raypeat.com
Eugene Oregon
USA

Third Edition.

Book Cover Image: "Glad Day"
(or, "The Dance of Albion") by William Blake.
Blake created and produced several versions of this etching on copperplate using a method he invented and called "Illuminated Printing" circa 1794-1810. The original work is held in the British Museum in London, other copies are held by the National Gallery of Art in Washington, D.C. as well as other institutions.

Ray Peat® Publishing
PO Box 5764
Eugene OR 97405
www.raypeat.com

CONTENTS

INTRODUCTION

In this book I offer some practical methods for solving health problems by increasing biological energy. The methods can be applied by people to themselves, or by physicians to their patients. But energy and individuality and thought are so closely associated that full access to the value of these techniques might require the integrated and balanced approach that can be provided by a new theoretical orientation. I offer some suggestions on the way our inner world can mesh with the outer world, in such an energy-based orientation. And the theory might not be of general interest unless its meaning can be concretized in some specific situations. Other people are working on the same kinds of theories, and other people have worked out parts of the various techniques, but my effort here is to offer something like a "technology of life," which unites theory and practice.

I intend to show that the human substance isn't simply a metastable, transient event, but that it is rather a pioneering probe into a new system. Self-organizing systems decay only if they have assimilated inertia, and--with a little support of the right kind--the centers of degeneration can become centers of regeneration.

Biogenic stimulators, adaptogens, spontaneous healing, cancer, aging, and reproduction are considered here in the light of the best available knowledge of developmental systems in biology and non-equilibrium processes. Besides having the practical value of bringing coherence into our understanding of some things that have been merely "therapeutic approaches," I hope it will also help to shift the theoretical focus in the study of self-ordering processes away from the informational-statistical, and toward the energetic; to interpret structure, phase and state in terms of energy and history, before considering their abstract, informational properties.

PART I

ASPECTS OF WHOLENESS

"Reason, or the ratio of all we have already known, is not the same that it shall be when we know more. The bounded is loathed by its possessor. The same dull round, even of a universe, would soon become a mill with complicated wheels. If the many become the same as the few when possess'd, More! More! is the cry of a mistaken soul; less than All cannot satisfy Man."

W. Blake

Once we begin to believe in the future, we understand the possibility of learning more, and of being more. We appreciate the unexpected, and we even anticipate the opportunity to confront it.

There is an inertia that makes it easy to over-value present knowledge. If we have enough energy (and enough time), we overcome the inertia. The desire for wholeness can lead us toward a more appropriate kind of science, and also toward a more perfect world.

CHAPTER 1

ASPECTS OF WHOLENESS

Holism is the observation that, although natural objects can be resolved into their parts, the parts are to some extent shaped by their participation in the whole object. For example, in organic chemistry it is recognized that the reactivity of an atom or radical is modified by adjoining parts of the molecule, by the solvent, etc. Holism would never have been named, except that it is so common in our scientific tradition for someone to say "We know all there is to know about the parts (atoms, surfaces, fields, genes, etc.), so we can foresee the result of their combination." Good engineering involves knowing the properties of the materials so thoroughly that accurate predictions can be made about their behavior in new structures, but good science requires a willingness to accept the unpredicted when it occurs. Holism, or a non-dogmatic attitude toward the world, recognizes individual uniqueness, rather than averages, and is likely to look for complex causes (especially environmental influences), rather than too easily ascribing traits "to the genes."

Historically, a reluctance to distinguish our present knowledge from possible knowledge, and to distinguish our definition of something from its real existence and fullest potential, has characterized most of the people who oppose holism, and call themselves reductionists. Consciousness, perception, sensation, pleasure, and intention have often been omitted from the world described by reductionists. If we are going to understand life and its possibilities, then it seems that we should begin with an appreciation of its "liveliest" aspects, as an essential dimension of our thinking, even if we are going to work with some of its relatively inert aspects, such as viral genetics.

The relation of energy to structure is, I think, the central question of biology. (The importance of the same question for the physical sciences might indicate in a rough way how Haldane's predicted "swallowing up" might occur.) The ideas of resonance and hysteresis, which are only vaguely defined in physics, have to do with the interaction of energy and structure on various levels of complexity and organization, and are examples of physical concepts that can gain meaning and clarity from biology.[1] When energy flows through matter, order accumulates (as a result of resonance and hysteresis, for example), but we hear so much about "entropy," "randomness," and "symmetry" that we forget most of the formative processes in the material world.

Human (and ecological) health obviously should have the benefits of holistic science, but the actual situation is that biology and medicine have become very product-oriented, and holistic considerations are increasingly left to a variety of "fringe" occupations. Many of these alternative approaches are concerned with the idea of "energy" as the key to health, but in general they lack simple and effective methods for optimizing biological energy, and often use counterproductive methods, such as prescribing linseed oil nutritional supplements. In the following pages I will show how some of the most important achievements of ordinary science can be retrieved from the distortions of the medical promoters, and made available for holistic use, that is, for appropriate use.

NON-GENETIC BIOLOGY

The word, "gene," itself contains an ideology, since it implies origin, or genesis, though its main meaning is something like "a unit of continuity." Building on the word's connotation, dogmatic geneticists explicitly stated their "central dogma": that information flows only from DNA to RNA, and only from RNA to protein. When Temin and Baltimore described the "reverse transcription" of DNA from RNA, all of the professional biologists I talked to said flatly that they didn't believe it was possible. Even when no one was threatening the addition of "information" to the chromosomes, the dominant belief among biologists was that development was closely controlled "by the genes," because the body owes its genesis to the genes. Intelligence, body proportions, and senescence were said to be "specified" or "governed" or "programmed" by the genes. The "congenital" condition was often taken as the "genetic" condition. Textbooks said that the maternal influence was only "genetic," because the fetus was "insulated" from events in the mother's body (just as the "germ line" was isolated from events in the rest of the body). Although many kinds of experiment showed both prenatal and transgenerational (see Figure 1.1) influences of the environment on intelligence, body proportions, and rate of aging, the genetics-reductionist school ignored them, and defined themselves as the only scientific school of biology.

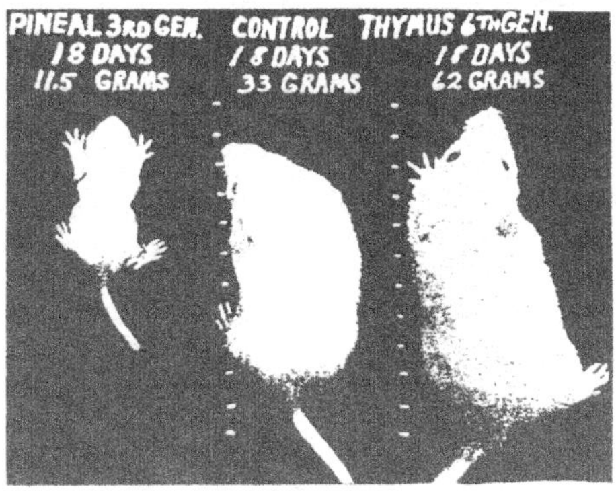

Figure 1.1 The cumulative effects of feeding glandular tissue.

ENERGY AND THE BRAIN

When I was a student of literature, I was interested in the apparent energy differences between different writers and different social groups. The literature of dissent (I wrote my master's thesis on William Blake) always reveals an abundance, or excess, of energy. I came to see literary "periods" or styles (classical, realist, Romantic, surrealist, etc.) as reflections of a society's energy and structure. As I followed this line of thinking about the function of language in society, and in the life of the individual, I saw that Blake's descriptions of his own use of images and symbols made sense in relation to brain physiology. Even attitudes toward mathematical concepts, theories of ideal languages, and "language philosophy" related strongly to the person's attitude toward authority, energy, the body, and the nature of consciousness.[2]

After I had spent a few years studying and teaching linguistics, I decided to study brain physiology. At that time, a group at the University of California had already found that freedom and stimulation caused a cumulative, transgenerational increase in the size of rats' brains. Continuing that work, Marion Diamond has found that pregnancy, or progesterone treatment increases brain size and inhibits the development of age pigment (suggesting that longevity might also be increased). In 1968, another group found that prenatal hormone treatment (causing increased production of progesterone) greatly increased the intelligence of rats. Several researchers have shown that brain size relates strongly to both energy production and longevity. A graph (Figure 1.2) suggests how adaptive (mental) energy and longevity might relate.

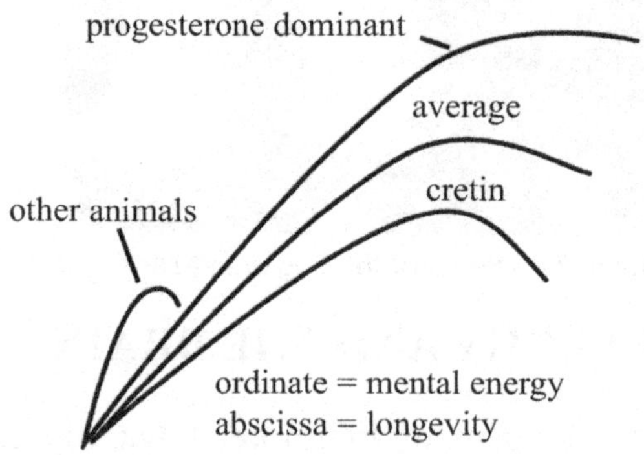

Figure 1.2 Graph suggesting relationship between adaptive (mental) energy and longevity.

At a certain stage of gestation, there are more than twice as many brain cells as there are at the time of birth, and it is likely that good conditions for the mother and fetus--adequate blood volume, oxygenation, glucose, progesterone[3], thyroid hormone, and absence of toxins and nutritional imbalances--can preserve some of the cells that would normally die. Since the brain and the liver regulate the amount of glucose and hormones in the body, the conditions under which those organs develop will very likely influence their later functioning, and account for some of the transgenerational effects that have been observed.

MECHANISMS OF AGING

As I became aware of the influences of hormones and the availability of energy on the development of the brain, and the effects of the brain on aging (and became aware of the amount of dogma in the neurosciences), I began to work on the ways in which energy metabolism affects reproductive failure, especially in senescent decline of fertility. I hoped that working in a less dogmatic area of physiology would still allow me to get a better understanding of the organism in general, including the brain.

Professor Soderwall and his students at the University of Oregon had shown that the corpora lutea (areas in the ovary which mainly produce progesterone) appeared to fail in aging hamsters, and that vitamin E supplements could extend fertility by a significant amount. His group showed that "aged ova" were not responsible for infertility, but rather that the uterine environment was not suitable for implantation. Soderwall had also demonstrated that excess estrogen could cause failure of the pregnancy at any point, from failure of the embryo to implant, to resorption of the fetus at a late stage of pregnancy.

Although I had investigated the association of estrogen with cancer, and knew from my own experience with migraines that stress, diet and hormones interacted in powerful ways, when I began to investigate the oxidative metabolism of the uterus I didn't realize that it would involve a convergence of several of my main interests. I was familiar with Otto Warburg's famous idea that cancer is caused by a "respiratory defect," and I knew that aged tissue has a diminished respiratory capacity.

The textbooks indicated that estrogen deficiency and "aged ova" were responsible for senescent infertility. (The origin of these ideas in a dogmatic matrix is another story which I have told elsewhere.) I found that the uterine endometrium of old animals often consumed oxygen at a high rate, and showed other signs of being under the influence of excessive estrogen. As I tried to understand this, I saw that several things could contribute to a high rate of oxygen consumption. Either too much estrogen, or too little progesterone could have the same effect, since it is the ratio between these hormones which controls their effects. A vitamin E deficiency increases oxygen consumption, and too much unsaturated fat has the same effect. In a vitamin E deficiency, unsaturated oils are oxidized in a way that produces "age pigment," also called ceroid pigment or lipofuscin. This pigment consumes both oxygen and fuel, but produces no usable energy. Estrogen excess synergizes with a deficiency of vitamin E to intensify the formation of this pigment. Partly, this might be because estrogen is a powerful stimulant of iron absorption, and iron is involved in the peroxidation that produces the pigment. But low oxygen concentration is what causes the iron to become active in peroxidation, and estrogen acts in several ways to decrease the availability of oxygen. It appeared, therefore, that many of the features of aging resembled an estrogen excess, rather than estrogen deficiency.

The way in which estrogen prevents or terminates pregnancy seems to be by causing the uterus to consume oxygen at such a high rate that there is no oxygen available for the embryo, which has a high requirement for oxygen beginning on the day that it normally implants. The chronic or cumulative effects of estrogen, leading to formation of lipofuscin, happen to act in the same direction as estrogen itself, causing oxygen to be reduced, especially in the uterus, but in all other tissues, too (aging the whole organism). Estrogen excess can also destroy the corpora lutea, interfering with the production of progesterone. Progesterone's effect in pregnancy is to assure the availability of oxygen and nutrients for the embryo, but it also has the general effect of inhibiting the formation of lipofuscin, and of other aging signs, by improving efficiency. (Progesterone is unusual among the anti-stress steroids in having no harmful side effects.)

Although my work confirmed the other research that had been done in Soderwall's lab in the preceding 25 years, the idea that estrogen's influence appears to increase with aging, and even to contribute to the process of aging, was contrary to the doctrine that has been promoted by the pharmaceutical industry. Nevertheless, as I read more, I saw that there was really no evidence contrary to what I had seen in my own work. What existed was a web of interpretation which existed to sell estrogen treatments. Even the fact that estrogen causes abortion was "ignored," very consciously, until the industry had fabricated a more acceptable rationale with which to sell its "contraceptive" pills.

The idea of many factors acting in the same direction, and tending to have a cumulative effect, seemed to me to have a general biological significance. It seemed to be part of the answer to the question of what it is which is lost, or accumulated, during aging, which accounts for the decreased ability to adapt to the changing environment. It seemed to say something about the nature of Warburg's "respiratory defect" in cancer. The thyroid hormone, which governs respiration, is suppressed by estrogen, and by unsaturated fats. The wasteful metabolism of estrogen dominance tends to use up glucose, and is likely to activate the stress hormones, including cortisol. The main features of aging can be produced directly by administering excessive amounts of cortisol. These features include atrophy of skin, arteries, muscle, bone, immune system, and parts of the brain, loss of pigment (melanin), deposition of fat in certain areas, and slowed conduction velocity of nerves. The physiology of aging (especially reproductive aging) overlaps the physiology of stress.

LIFE EXPECTANCY OF ADULTS

When there are sewers and clean water, the death rate of infants and children from infectious diseases isn't high, so the average age of the population increases as a larger percentage of those born live into old age. Increased use of birth control will also increase the average age of the population if fewer children are born each year. These changes in the birth rate and infant survival can happen at the same time that industrial pollution is causing chronic poisoning which shortens the lives of mature people.

To understand the effects of aging on the risk of human disease and death, we need information specifically about people who have reached a certain age. There are several organizations and individuals in the U.S. that have cleverly used "re-standardized" statistics to convince the public that cancer is being defeated, industrial pollution is harmless, sickness is caused by the victim's own irresponsible choices, and that our longevity has increased by several years in the last few decades. Old people are the most likely to die of cancer, and the cancer death rate for old people in the U.S. has been increasing in recent decades, as has the cancer death rate for the population in general. The life expectancy for older people in the U.S. has barely increased a year in recent decades. 2,000 years ago, the life expectancy for older people seems to have been several years longer than it is now (Figure 1.3).

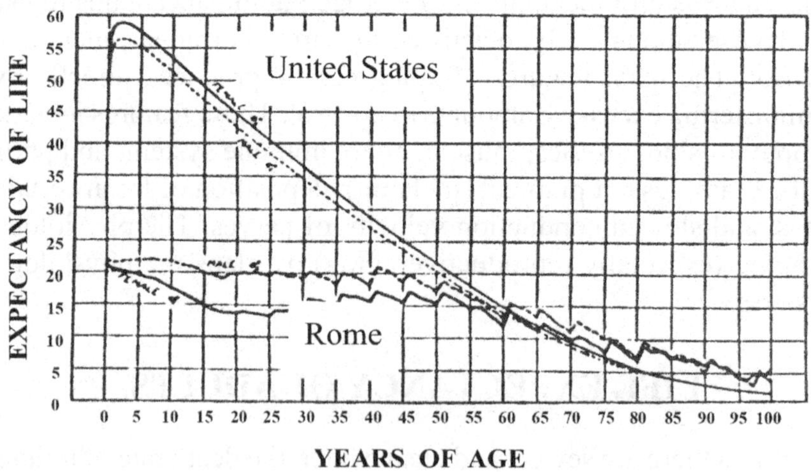

Figure 1.3 Comparing the expectation of life of Ancient Romans with that of present day Americans. Plotted from Macdonell's and Glover's data.

PUBERTY AND AGING

Many studies have demonstrated that puberty seems to trigger the mechanism of aging, and the idea of a "death hormone," located in the pituitary gland, has been suggested. (Figures 1.4 and 1.5) Many degenerative diseases develop under the influence of excessive estrogen and cortisone (and as a result of the many metabolic changes which follow exposure to those hormones). Many of these diseases, especially those which appear after puberty and are more frequent in women, can be treated very effectively with the anti-estrogen and antistress hormones, such as progesterone.

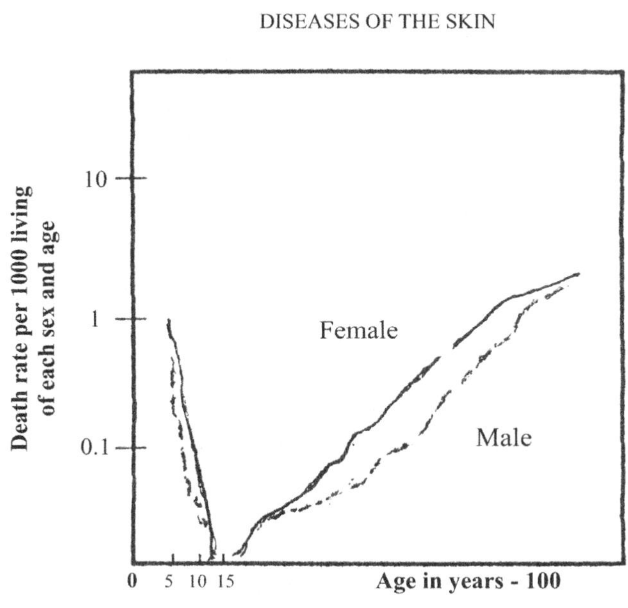

Figure 1.4 The shape of this curve is exactly like that for death from cancer, death from all causes, and concentration of iron in the tissues.

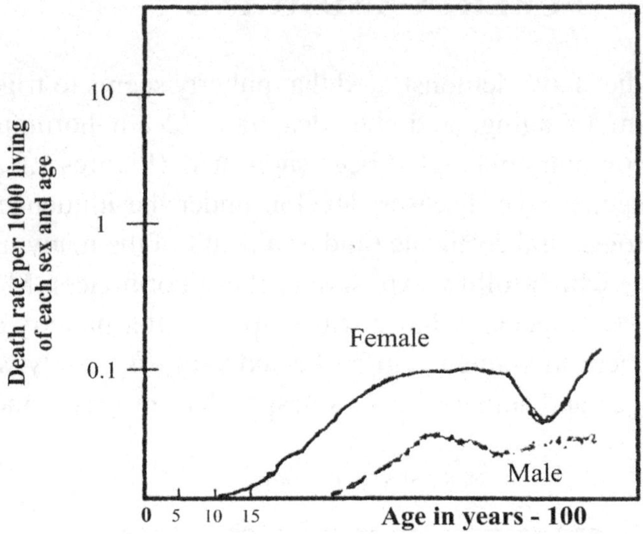

Figure 1.5 Endocrine Diseases: Impact on mortality by gender and age.

Hans Selye pointed out that estrogen treatment mimics the first, "shock" phase of the stress reaction. An excess of estrogen (or any stressor) causes the pituitary to secrete prolactin and ACTH, and both of these hormones act on the ovaries to stop progesterone production, and contribute in many other ways to the process of atrophy. ACTH, of course, stimulates the secretion of cortisol. The removal of the pituitary obviously isn't a practical way to delay senescence, but protection against the "death hormones" can be achieved to some extent by altering the diet to minimize the effects of estrogen and cortisol.

Historical and demographic studies show that certain conditions affect the age at which puberty occurs. Ashley Montague has argued that we need more neoteny, that is, that we should try to preserve and to extend our youthful functions, because those are our most human qualities. If we can generalize from animal studies, delaying puberty could increase brain size and longevity, improve intelligence, decrease violence, and even make people physically more attractive (the "cute puppy" appearance is largely a matter of brain size in relation to the size of the face and body). I think this will be the next step in human evolution. Just as nurturing, stimulation, and freedom promote improvement in the function and structure of the brain, cruelty and oppression act in the opposite direction. If puberty is delayed, then the importance of a culture which supports curiosity, exploration, play and sensuous pleasure seems obvious, so that biology and culture will be mutually supportive.

The physiological age of the parts of an organism depends in some way on the developmental stage of the whole organism. This contradicts the reductionists' idea that cells or tissues have an "intrinsic" lifespan which will cause them to deteriorate after a certain limited number of divisions. When pieces of breast tissue or skin were repeatedly transplanted from old animals to young animals of the same (syngeneic) strain, they were still in good condition after ten "lifetimes," and their survival was apparently limited only by the necessity of trimming them each time they were transferred, to make sure that no host tissue was transplanted with them. When old rats were grafted onto young rats, the old member of the pair lived to twice the expected age. Recently, young female mice were grafted onto old females, to investigate any hormonal factors in aging. The ovaries of the young animal appeared to age, and its production of progesterone decreased. That is, old tissue seems to become younger on young animals, and young tissue can be "aged" by association with an old animal.

This kind of evidence (and the simple observation that the cells in skin and intestine undergo thousands of divisions in an individual's lifetime) strongly favors the idea that a systemic energy problem is involved in aging.

ENERGY AND EVOLUTION

V. I. Vernadsky believed that the earth's energy-exchange and substance-exchange processes were intensifying, and that the biosphere would undergo another major increase in its "metabolic rate," similar to that which appeared early in life's development with photosynthesis. M. I. Budyko (in *Evolution of the Biosphere*, 1986) discusses the principle of "aromorphosis," in which the origin of higher types of animal is closely associated with the availability of larger amounts of energy in the environment, and with the appearance of new structures which made possible a higher energy level of activity and more complex interactions with the environment.[4] When vertebrates developed an effective method of heat regulation, they were able to considerably increase their use of energy. The move from water to land requires greater ability to use energy, as well as requiring new structures suitable to the new way of living.

When mammals and birds achieved the ability to sustain a high metabolic rate by keeping their bodies at a steady, fairly high, temperature, their "food chain," based on photosynthesis, consisted of organisms that generally lived at a lower temperature.

Sugars, proteins, and the saturated fats produced by warm organisms can be eaten by warm-blooded animals with no particular side effects.[5] Organisms that live at low temperatures, however, contain unsaturated fats. The consumption of large amounts of unsaturated fats lowers the metabolic rate, and accumulated unsaturated fats are susceptible to a spontaneous and toxic form of oxidation. (The toxic effects include damage to the respiratory apparatus and to the circulatory and immune systems, increased rate of aging, and cancer.)

These "low energy" foods in effect counteract the evolutionary achievement of a high metabolic rate. Several studies show that decreased consumption of unsaturated fats can delay puberty.

Other studies show that an excess of unsaturated oil in the mother's diet can damage the development of the fetus's brain. The choice of foods which have less unsaturated fat tends to reinforce the achievements of evolution.

The seed oil industry has created a national phobia about the consumption of saturated fats and cholesterol, but there is no basis for the idea that those foods should be avoided. People with hypothyroidism are susceptible to heart disease, but their elevated blood cholesterol becomes normal when their thyroid function is restored. (Figure 1.6) The body's highest concentration of cholesterol exists in the brain. The level of cholesterol in the blood strongly influences the production of the protective hormones, such as progesterone. The brain contains by far the body's highest concentration of these hormones.

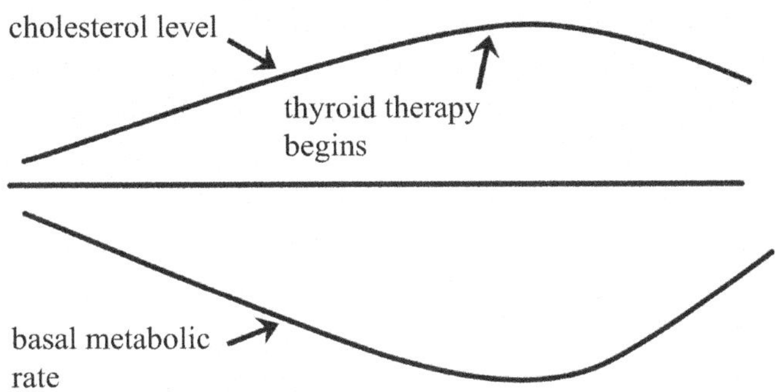

Figure 1.6 Elevated blood cholesterol tends to normalize as thyroid function is restored.

Another energy-promoting factor is visible light. In the winter and at night the respiratory energy producing system is damaged, and the protective hormones decline, and the harmful stress hormones increase. The immune system becomes, less active, and mortality increases.

Although ultraviolet light interacts with unsaturated fats in the skin to accelerate aging (E. R. Pinckney, *Medical Counterpoint*, Feb, 1973) and to produce cancer, ordinary visible light has several beneficial effects in animals. One effect is the "regeneration" of the enzyme SOD (superoxide dismutase), by causing its copper atom to be reattached to the protein. Light also increases the activity of normal respiratory enzymes, and tends to normalize (or maximize) the production of hormones, including progesterone and thyroid.

Animal migration to reproduce in regions with longer days is a way to benefit from this energy-promoting action of light. In adult birds, the increase of hormones in the spring causes the growth of new brain cells in the area that controls their singing. (In humans, the space inside the cranium keeps increasing into old age, and the amount of DNA in the brain also keeps increasing with age, but it has been assumed that such changes in adult brains result from an increase in the size of nerve cells, and an increase in the number of connective tissue cells, rather than from a continuing increase in the number of nerve cells.) I would expect an increase in the temperature of the earth, and increased use of artificial light (or migration) to lead to a prolongation of youth and the development of better brains.

Many years ago, people talked about the tendency toward "cephalization" in evolution. That is the tendency to create a hierarchical organization, with the size and importance of the brain increasing with evolution. More recently, it was found that both the intensity and the efficiency of respiratory energy production in the brain increase with evolutionary level and with the degree of alertness.

Another seemingly directional tendency relating to energy production has to do with a non-random composition of the DNA. Aerobic microorganisms' DNA has a higher G+C (the bases guanine and cytosine) content than does that of anaerobes, and warm-blooded vertebrates have more G+C than do lower organisms, with the G+C content again corresponding to the amount of oxygen used. Although the full meaning of the G+C content isn't known, it is thought to affect the stability of the organism at various levels, and I suppose this would allow the organism to thrive at higher temperatures.

The association of G+C content with evolutionary level is interesting because the changes occur in "non-coding" DNA, where the ordinary kind of selection, involving "coded" proteins, isn't involved. The "isolation" of the genome, and the randomness of its changes, used to be at the core of genetics; I suspect this is why even most biologists aren't aware of this odd feature of DNA composition. The passage of time will allow people to disengage themselves from an embarrassing dogma. (Figure 1.7 shows the ascending G+C content for fruit fly, toad, mouse, cow, and human, in the DNA of the mitochondrion, where energy is produced.)

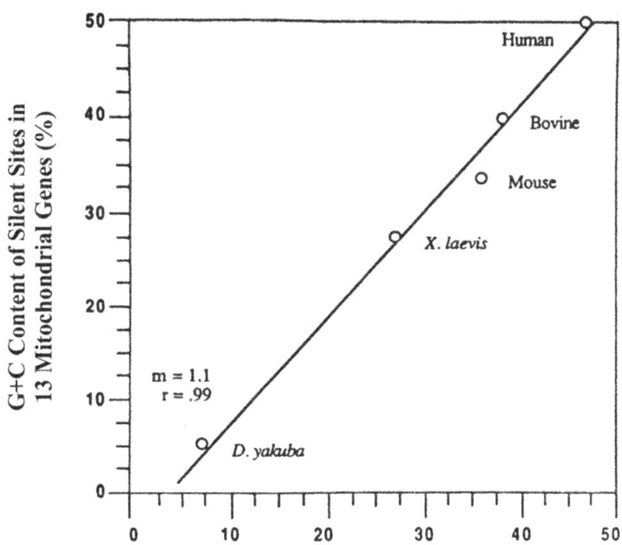

Figure 1.7 Jukes, T. H., & Bhushan, V. (1966). Silent Nucleotide Substitution and G+C Content of some Mitochondrial and Bacterial Genes. *Journal of Molecular Biology 24*, 39-44.

POLLUTION

Besides the distortion of our food supply by the propaganda of the seed oil industry, there is increasing contamination of our food supply by heavy metals. Lead, for instance, has been spread everywhere, largely as a result of its use in leaded gasoline. Food additives are often contaminated with heavy metals from the sulfuric acid used in their manufacture. Many people are aware of the famous experiments in which food restriction increased the longevity of animals, as if eating were toxic. But removing toxic heavy metals from the food, without restricting the amount of food eaten, has had the same life-extending effect in experimental animals.

When estrogen is elevated (as at puberty, or in pregnancy, or with medical estrogen treatment) the absorption of iron is stimulated. During aging, the body's load of iron increases, especially if there is a deficiency of copper, and the body's content of copper decreases with age. Copper is an essential component of cytochrome oxidase, which has the crucial last position in the mitochondrial respiratory system. Copper is a component of the cytoplasmic SOD enzyme, which decreases with age. Ceruloplasmin, a major copper-containing protein, helps to keep iron in its safe oxidized form. Copper is involved in the production of melanin (itself an antioxidant) and elastin. The loss of melanin, elastin, and respiratory capacity, which is so characteristic of senescence, is also produced by excessive exposure to cortisol.

The protein, metallothionein, is rich in sulfhydryl groups, which bind heavy metals. It is assumed that this protein helps to detoxify and eliminate the toxic metals. This protein is induced by exposure to either a heavy metal or cortisol. A larger amount is produced in response to the combination of heavy metals and cortisol, than by either alone. Since copper, like the toxic metals cadmium, lead, mercury, and silver, reacts strongly with sulfhydryl groups, the body is likely to lose some copper when it is subjected to heavy metals or cortisol. I think this chronic loss of copper accounts for the obvious features of aging, such as loss of elasticity of the skin, lungs, and blood vessels, the depigmentation (demelanization) of skin, hair, and (in Parkinson's disease) substantia nigra, and for the decrease in respiratory capacity. The replacement of the copper by iron (and the loss of the copper enzymes which protect against iron-catalyzed free radicals) probably accounts for the increased formation of lipofuscin during aging. When copper-dependent mitochondrial respiration fails, lipofuscin has the ability to sustain energy production through glycolysis (by keeping the coenzyme NAD, nicotinamide adenine dinucleotide, relatively oxidized), so it is possible that lipofuscin is a primitive sort of defense against stress.

In animals, copper supplementation can restore natural color to white hair, and in one experiment, it increased longevity. At present, there isn't enough knowledge about the safety of different ways of administering supplemental copper. It can be toxic, and it oxidizes other nutrients. Besides choosing foods high in copper and low in iron and other heavy metals, other dietary choices which support thyroid function will tend to promote the retention of copper. Other dietary practices can minimize our production of cortisol (e.g., combining fruits and proteins, since protein foods lower blood sugar and stimulate the secretion of cortisol).

SELF-ORDERING PROCESSES

An old idea in our culture, which still inclines many people to a "billiard ball" conception of mechanistic causality, is that matter must be ordered "from outside." For Descartes, this involved a material world which lacks "secondary" qualities, i.e., every quality other than being present in space. That abstract world was supplemented by an utterly separate world of mind and spirit.

Descartes' spirit has continued to live in many geneticists, who see "selection" as an ordering principle imposed upon a stupid sort of material substance, which can vary only randomly, by chance.

This genetic doctrine has a strong influence on medical thinking. Hundreds of diseases have been classified as "genetic diseases," and the idea is extended by showing an increased susceptibility among certain genetic tissue types even to germ-caused diseases and toxic sicknesses. Diabetes, allergies, arthritis, goiter, and myopia are often said to be "caused by genes," despite clear evidence for the involvement of the environment in producing them.

Several famous molecular biologists have been writing and lecturing about their idea that everything in our behavior and culture (including art, politics, and language) is specifically programmed by genes. (It would seem that I performed genetic surgery on one of these people, or caused him to mutate, when I pointed out some glaring errors in his argument, and saw that he subsequently changed his claims.) Physical science provides a much richer picture of the qualities and potential of the material world than geneticists recognize. Even many physicists don't recognize the richness implied by the body of experimental results in their field. Many well known physical scientists have had relatively holistic attitudes (e.g., J. C. Bose, Michael Polanyi, B. V. Deryagin, Frederick Soddy, V. I. Vernadsky). A rich view of physics has much to offer to biology. However, when I say that a holistic view of biology is open to using physics and chemistry, as well as ecology, history, and cosmology, to achieve an adequate understanding, I should mention that there is a school of weird (immaterial and "quantum" centered) physics which is presenting itself as a holistic world view. To them, I think Einstein's remark still applies: "You believe in a dice-playing God and I in perfect laws in the world of things existing as real objects. . . ." Elsewhere, Einstein observed that an object's fields amounted to an extension of its material substance, i.e., he preferred to materialize fields, rather than to dematerialize things, as some of the popular philosophers of physics do.[6]

The orderly, epitaxial growth of crystals has been shown to occur across the thickness of a plastic film. A detailed study of this sort of long-range ordering process was made by Alexander Rothen. He was able to demonstrate biologically specific adsorption at relatively great distances. Many other types of research in adsorption fields and long-range order make it clear that the interactions of atoms and molecules in cells needn't be governed by either direct contact or by random motion. When cell components are rearranged, they return to their normal position in relation to other components, revealing a great capacity for self-assembly or self-ordering.

The medical tradition of naturopathy recognizes a great capacity of the body for self-regulation and self-healing. I think these attitudes can be usefully expanded now, in the light of new knowledge about energy and structure. On the short time scale in which we think about the health of an individual, and on the transgenerational scale relating to having healthier, more intelligent children, and on the evolutionary time scale, I think we can see a tendency, not just to preserve homeostasis, but to move upward in energy and greater generality of structure and function. To provide more energy and scope for using it stimulates our ability to use energy meaningfully.

SOME IMPLICATIONS

There is considerable flexibility in living organisms, and in higher and lower levels of organization, and we can see some of the ways in which structures of different complexity accommodate themselves to the surrounding condition of energy and structure.

The conditions under which the brain develops, including gestational support and the later nutritional, hormonal, and behavioral conditions, the degree of stress and stimulation, contribute to the brain's structural complexity and metabolic energy use, and to the organism's ability to cope successfully with the environment.

Vicious circles of physiology often stabilize an organism on a low energy level, which may involve disease or rapid aging.

The existence of a few systems of positive feedback (self stimulation), however, indicates that in our fundamental structure we are biased in an expansive, upward direction. Progesterone (and its precursors, pregnenolone and cholesterol) and thyroid hormones participate in some of the important positive feedback systems, involving energy production, stress resistance, and brain growth.

In therapy and in everyday living we can try to protect and promote our energy-producing and energy-using systems by seeking the stimulation, the conditions of light and temperature, and the foods that are appropriate for our evolutionary level.

NOTES

1. Resonance has to do with approximately simultaneous interactions between energy and matter, and hysteresis, with the lingering effects of earlier interactions of energy and matter.

2. I circulated a questionnaire in 1957 among college students of various nationalities and classes. My conclusions suggested that the contemporary white male Western abstract personality was dysadaptive, in the sense that many vital abilities or capacities were undeveloped or suppressed.

3. Progesterone is one of the factors in a "chemically defined" medium for the growth of nerve cells in culture. The others are insulin, transferrin, selenite, and putrescine.

4. Budyko's idea that energy drives the biosphere is unfamiliar to our Anglo-American culture, which has tried to think of everything, even the weather (and especially evolution) in terms of mechanics, inertia, statistics, and hydrodynamics. There is a resistance to the idea of directional, dynamic self-governing systems. The sun's energy for photosynthesis tends to be kept in a background role.

5. "Warm organisms" means either organisms that keep warm by internal heat production and temperature regulation or organisms that are warm because they are kept warm by the environment.

6. Paul Forman, in an article in "Studies in the History of Science," gives some valuable historical information on the social factors involved in Germany of the 1920's, when physics' departure from the world of real objects became so conspicuous.

CHAPTER 2

ANOTHER VIEW OF EVOLUTION

A few people have pointed out that order can't be expected to emerge from disorder, even "if you wait long enough," unless an ordering principle, a "formative tendency," is present in the starting situation. "Chaos," then, would be described as "potential," and the development of organization would be more like a snowball rolling downhill, than like the step-by-step preservation of "random variations" over an almost infinitely long time scale. Our knowledge of ourselves, and our sense of the meaning of life, are influenced by the answer to this basic question about the nature of order, and of the origin of life.

When complex organic compounds were found in petroleum, some people argued that this was evidence that the petroleum was the residue of dead organisms, that had seeped into porous geological formations. Another view is that petroleum was directly created by the earth, and that life is a parallel creation.

In 1859, the year Darwin published his famous book, the first oil well was drilled in Pennsylvania, and chemists began to study petroleum with increased interest. When high-carbon iron is dissolved in hydrochloric acid, an oil appears on the surface of the acid. Mendeleev noticed that it resembled natural petroleum, so he gave samples to people who were familiar with petroleum from different parts of the world. All identified its smell as that of petroleum, and some even thought they could identify the region it came from.[1] Mendeleev and others proposed that molten metals deep in the earth might interact with the carbon present in minerals to create petroleum.

Recently, Gustaf Arrhenius at the Scripps Institute of Oceanography has been studying the ability of iron carbonates to produce organic compounds. I think the gap of 100 years between this work and the ideas of Mendeleev and his contemporaries is the result of a dominant habit of thought which sees the world as an inhospitable place, in which life is a great improbability that came into existence only with great difficulty. The famous Crick, for example, suggested that life had to originate elsewhere, and arrived on earth as a truly alien space-traveler. The thought that life forms might just sort of gush up out of the earth in volcanic regions makes it all seem too easy; where might it lead if people started believing that life could originate without a struggle for existence?

Several years ago, Sidney Fox demonstrated that amino acids placed on a hot volcanic rock spontaneously formed into proteins, and that these proteins had some enzyme-like catalytic action. Adding a little water, the proteins spontaneously formed small self-reproducing spheres, just the size of bacteria. One of their enzyme-like actions was the ability to synthesize gene-like chains from suitable precursor molecules.

Explorations of the deep ocean floor, far below the level at which life should exist, because of the absence of light, revealed worm-like organisms. Further study showed that the worms live on bacteria, which in turn live on volcanic emissions rising through the ocean floor. If volcanic processes created life forms, and can even provide their food, then we don't have to imagine that green plants necessarily evolved before animals.

Oxidants[2] are emitted by volcanism, and there are heme-like porphyrin molecules in petroleum which can catalyze respiration-like reactions. While molecular oxygen is now our most important electron acceptor (oxidant), other substances could have had a similar function. And, since volcanos emit both oxidants and reductants (depending on temperature and other conditions) we shouldn't blindly accept the doctrine that life originated in a "reducing atmosphere."[3] Since the reactions which produce petroleum-like hydrocarbons do not produce carbohydrates, we might want to investigate the possibility that carbohydrates are the product of a primitive type of metabolism.

Several things look different when we consider the possibility of a volcanic origin for life. For example, there is a suggestion that life originated in abundance, in an environment rich, at least locally, in complex organic molecules, rather than in occasional puffs of molecules generated by lightning or dissolving slowly out of a meteorite. And the thought that life is truly geological, and is an expression of our planet's nature, might make us feel more at home, and nurtured.

If we think of life, already complex and respiring, moving into the light and adapting to a new system in which carbon dioxide is captured from the environment for new syntheses, we are likely to ask a different set of biochemical and bioenergetic questions. Detoxifying systems, that would have evolved to protect respiration from compounds such as ammonia, cyanide, and carbon monoxide, might have been the first systems to make use of light. Evolution of these systems might have led into photosynthesis. If this is true, then the theoretical and chemical study of the origin of life might give us some clues to the nature of the biological rhythms based on the cycle of day and night.

Several people have written about the idea of a "noosphere." Pierre Teilhard de Chardin, in *The Human Phenomenon*, used the term coined by Edouard Le Roy. V. I. Vernadsky, in *Biogeochemical Essays*, developed the concept in relation to the principles of biogeochemistry. Vernadsky saw the noosphere as a stage in the natural geological development of the earth. He suggested that the various stages and aspects of the earth, the lithosphere, the hydrosphere, the atmosphere, the biosphere, will be followed by a stage in which a highly developed human society will recognize its place in the world and with understanding will guide the preservation and development of the natural system.

This is not exactly the same as the "Gaia hypothesis," which sees the earth as an organism-like self-regulating system, since the people who talk about the noosphere idea recognize the damage that human life has already done to the biosphere, and emphasize that the course of human civilization will be crucial in the fate of the earth.

If we proceed blindly with aimless industrialization, deforestation, destructive agricultural practices and militarism, the entire biosphere could be destroyed, possibly ending the earth's progress forever. But if civilization proceeds toward an understanding that the world is in development, we might find our place in this developmental process. If this happens, the earth will be entering a new energy level, analogous to the earlier appearance of photosynthesis. As we build our understanding we can increasingly think of our consciousness as a geological force.

During the Second World War, Vernadsky wrote: "I look at everything from the point of view of the noosphere and I think that in the wind and storm, in the horror and ordeal, a wonderful new future is being spontaneously born."

NOTES

1. Kropotkin, P. N., (1986). D. I. Mendeleev's hypothesis on the inorganic origin of petroleum and its modern scientific development. *Mendeleev Chemistry Journal, 31*(5), 1.

2. Many of the conditions that exist in volcanos are able to generate molecular oxygen, e.g., heating metal oxides, oxyacids and oxy-salts, adding chlorine to a lime solution, or combining chlorine and steam at high temperature. If released in explosions, the molecular oxygen could be overlooked.

3. The "argument" that life originated in a reducing atmosphere is that ultraviolet light causes that kind of atmosphere to form many organic compounds; oxygen would form ozone, blocking the ultraviolet radiation. If ultraviolet light created life, oxygen must have been absent. If something other than ultraviolet light produced the complex organic molecules, the presence of oxygen would simply have had the same protective effect it now has.

CHAPTER 3

VERNADSKY'S HOLISTIC SCIENCE

Increasingly in the last decade, people in the Soviet Union who are interested in perestroika and an alternative world view have been thinking about V. I. Vernadsky and his work. Ecologists and conservationists in that country have found that his picture of the world offers an approach to the unity of life and its environment which is both practical and inspiring. While it initially sounds like a strange thing to say, Vernadsky's work made a real place for human consciousness and history in science by using simple physical concepts such as mass, energy, and rates of change, for approaching the biosphere. By examining the limits set by space and the flow of energy from the sun and the earth, Vernadsky was led to propose some simple laws of biogeochemistry, for example that the migration of chemical elements tends towards a maximum. Within this framework, he saw that the intensity of biological reproduction and metabolism and the degree of biological organization have orderly and lawful interrelations, and that human culture and consciousness participate in these natural relationships and processes. Vernadsky described his view as "cosmic realism," a phrase that had also been used by Mendeleev.

As a student, Vernadsky (1863-1945) was drawn toward both history and mathematics, and he read widely in philosophy and religion, including Plato, Aristotle, and Hinduism. After studying with Dokuchaev (who has been described as the pioneer of ecology because of his study of the interaction of biological and geological factors in the formation of soil), Vernadsky studied in Germany and France. It seems that he sought out the best minds and strongest personalities of the time, in many branches of knowledge. He knew the Curies, Hans Driesch (a student of Haeckel, the Darwinist; Driesch is known as a vitalist), Sechenov,

Le Chatelier, Mendeleev, and Tolstoy, and was able to integrate their perceptions into his view.

For example, he wrote "I think that the teaching of Tolstoy is much deeper than I first thought. This depth consists in his view that the basis of life is the search for truth, and that the real task is to tell this truth without any retreats," and he described Tolstoy as an accumulator of the energy of human consciousness.

Tolstoy's moral thought was paralleled by Vernadsky's perception of unity, in which humans, as part of a cycle of transformations, have an obligation to think and act for the good of the planet. The creation and preservation of culture and human consciousness are given a central place in Vernadsky's view, but not at the expense of anything else in the world: Consciousness is needed by all of humanity, for the good of the world.

In 1922 Vernadsky lectured in Paris on biogeochemistry and on the place of consciousness in the history of the earth. These ideas were summarized in his 1926 book *The Biosphere*. He tried to get support in the United States to start a laboratory for the study of biogeochemistry, but his friend E. S. Dana at Yale was apparently the only American interested in the subject. The chairman of the chemistry department at Stanford said that the subject "hardly deserves the attention he would give it," and the proposal was also rejected by the Carnegie Institution, the National Research Council, and the British Association for the Advancement of Science.

Although many of Vernadsky's concrete discoveries have been independently rediscovered in the United States, his name seldom appears in American books on geology or ecology. I think this is because of his basic belief that nature is not ruled by accident. "Earth's structure is a harmonious integration of parts that must be studied as an indivisible mechanism. . . .Creatures on earth are the fruit of a long and complicated mechanism in which it is known that fixed laws apply and chance does not exist."

The idealization of randomness in German and Anglo-American science makes it hard even now to discuss physical and biological approaches which emphasize order, lawfulness, and directional transformations. Like Le Chatelier's ideas in physical chemistry, which were very influential in Vernadsky's thinking, the idea of a principle of "stability" which leads spontaneously to increased complexity and order in a system, is a little too alien to be received easily in our culture. Ilya Prigogine's work in this area is probably changing the atmosphere a little.

Unfortunately, even when there are systematic observations of the appearance of structure out of relative disorder, there is in the United States a preference to attribute the appearance of order to "chaotic" processes, and to argue from the "logic of chaos" that the future is absolutely unforeseeable, and that the world is chaotic rather than lawful. (If there is any significant chaos in this new doctrine, I think it is primarily in the terminology and in the illogical conclusions of some of its proponents.)

To Vernadsky, the world is lawful. His view of life as an active mass, a form of matter which must be understood quantitatively, "a whirlwind of molecules," acting on its surroundings, led him to think in terms of the rate of migration of atoms through this active mass. With this view, inert atoms become active, while active atoms become inert, depending on their situation.

This idea of the "history of an atom" is perfectly natural from the "cosmic realist" point of view, but it leads to questions that just don't occur to a person who usually thinks of atoms as statistical and anonymous identities. His first biogeochemical law, that the migration of atoms tends toward a maximum, is obvious to anyone who thinks about it, and it is an obvious parallel to Le Chatelier's principle of equilibrium. Once that principle is accepted, and if life is seen as part of nature, then Vernadsky's second law, that evolution will produce species which maximize the migration of chemical elements, follows naturally.

But this requires thinking about the biosphere as a system which is driven by energy from the sun, and this context isn't habitual for evolutionists. Vernadsky said that his second biogeochemical principle "shows, in a manner as precise as the corresponding principles in mechanics and physical chemistry, the direction in which the processes of evolution must proceed, namely, in the direction of increasing consciousness and thought, and of forms having greater and greater influence on their surroundings."

Teilhard de Chardin got his idea of the Noosphere from Verndsky's work. Many people prefer to see the idea of consciousness' having a meaningful and central place in evolution as exclusively a religious idea. However, from Vernadsky's perspective it can be derived from seeing life in physical and chemical terms. To make the implications of the idea clearer, it might be useful to apply it to some contemporary scientific questions, which could range from the origin of life, to the optimization of farming and dietary practices.

CHAPTER 4

THE CENTRALITY OF ANTICIPATION

"One gathers that the period of youth among mammals is not essentially recapitulatory but rather adaptive and anticipatory."

P. E. Davidson, in *The Recapitulatory Theory and Human Infancy*, Columbia University Press, 1914.

Energy flows through all systems, and the flow of energy leaves a residue of structure. In the 1960's and early 1970's, I made many "model" systems of the chemistry of respiration, trying to get perspectives on the meaning of biological respiration.

One of my experiments, in 1971, involved ascorbic acid, hydrogen peroxide, iodine, and copper. Every few seconds, the mixture alternately changed color and emitted puffs of gas; this is called an oscillating reaction, or a Belousov-Zhabotinsky reaction. This made me realize that "structure building" can involve the dimension of time, in a way more interesting than the simple dissipation of energy over time. When starting at a point, an oscillating reaction of this sort can generate interesting ring and spiral structures in space.

I see the interaction between the flow of energy (e.g., between a sugar and oxygen) and the structure as one in which the flow is retarded by the structure, and used by the structure, in maintaining and complexifying itself (Another way of looking at this is that if energy can do something, it does something; and what it does is to build structure. The life structure is a kind of energy charge, but the important thing is the spontaneous nature of the interactions, in the presence of an energy supply.) The chemical nature of the energy source (and of the energy "sink," which is usually oxygen) influences the nature of the structure-building.

A point made by O. Warburg and A. Szent-Gyorgyi and others is that there is an important difference between the energy provided by glycolysis and that provided by mitochondrial respiration. They felt that glycolysis was a more primitive form of energy production, and supported only primitive function and cell division, while the more efficient respiration supported cell differentiation and complex functioning.

Warburg thought that localization of mitochondria within cells could explain the special effects of respiratory energy, while Szent-Gyorgyi suspected that respiration more directly held the structural proteins of the cell in a special electronic state. It is now recognized that the localization of mitochondria within a cell creates an energy (ATP) gradient within the cell. (There has been a dogmatic disbelief in such gradients.) And, for example, the polyamines, which are associated with an increased rate of cell division, appear to shift cell metabolism away from respiration, toward glycolysis. The nature of the energy source seems to be more important than many biologists have thought.

Various hormones involved in adaptation and reproduction modulate the balance between glycolysis and respiration. Elsewhere (e.g., *J. Orthomol. Psych.*, 1975) I have discussed the hormone-like properties of various nutrients. In the 1930's, it was pointed out that all of the changes produced in rats by a diet lacking the "essential fatty acids" were the same as the changes occurring in hyperthyroidism. Parallel to the balance (or antagonism) between opposing hormones, there seems to be some sort of balance and opposition between different sorts of food energy.

The structural and functional repercussions of different dietary energy sources are complex, and I plan to discuss some of them soon in more detail, but there is some evidence for a relatively simple sort of absorptive or solubility compatibility between the cell substance in a given energy state, and the fuel molecules preferred by that state. Conceivably, this could account for the apparent "directed mutations" in the ability of microorganisms to use new substances for foods, observed by C. C. Lindegren, J. Cairns, and others. In adaptation to foods which are actually present, no anticipation is required, but only a sort of structural accommodation.

When the availability of food changes according to a daily cycle of light and dark, there is an adjustment of the metabolism involving many hormones, and an adjustment of nervous function, to protect against stress (Figure 4.1). Learning is involved, in the sense that a cyclic repetition is anticipated even by "simple" organisms. In humans and other mammals, there is the "dawn phenomenon," in which blood sugar rises to daytime levels before sunrise.

Figure 4.1 Blood cortisol rising during the night.

When food is abundant during pregnancy, and when the environment is not stressful, the brain of the fetus grows at a faster rate than in cases in which the pregnancy is affected by malnutrition or stress. Since the brain is a metabolically expensive organ, and since a larger brain tends to correspond to a higher metabolic rate of the entire organism, it appears that the developing fetus is anticipating a lifetime of relative abundance of food when it is well nourished in the uterus, and when it is exposed to less cortisol and more progesterone.

In evolution, the tendency toward dominance of the head (cephalization) in animals overlaps with another tendency (known in plants too) called juvenilization, pedomorphism, or neoteny, in which an early stage of the organism's development, the juvenile stage, is preserved for longer and longer periods in the descendants, eventually becoming the normal adult type. Baby apes resemble humans, in body proportions and behavior, much more than the adult apes do. The infant represents our evolutionary future.

Embryologists have observed that "ontogeny recapitulates phylogeny," that is, that early in development, mammals go through stages that resemble fish and reptiles. But I have always been intrigued by the fact that the same trend which can be seen going back from the adult to the infant--i.e., increasing of the brain-to-body ratio--can be seen to continue from the infant back to the embryo, at about the second month or seventh week of development. That is, the first few weeks of ontogeny recapitulate phylogeny, but then ontogeny also *anticipates* a phylogeny that hasn't been yet (Figure 4.2).

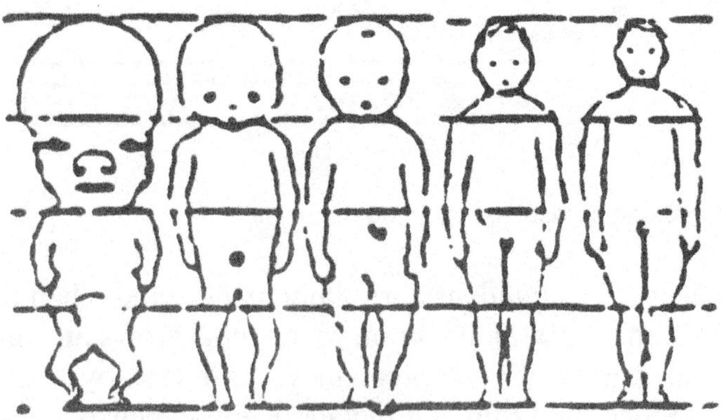

Figure 4.2 Ontogeny may anticipate phylogeny.

If the physical/chemical structures which retard the flow of energy, and complexify with it, accommodating themselves to it, reflect past and present conditions and types of energy source and energy sink, then presumably different geological phases would be suitably associated with different anatomies and physiologies. The extremely favorable energy relationships which exist in the early embryo (which is very small in relation to its supply of energy) appear to support a certain structural relationship, which sketches out the structural-functional possibilities for the future, when our environmental energy resources might be richer.

NOTES

The term "aromorphosis" has been applied to the sudden appearance of a new biological form, when energetic conditions are suitable (e.g., M. I. Budyko, *Evolution of the Biosphere*, 1986). Pedomorphosis, or juvenilization, can be thought of as the "aromorphic" response to an environment (or niche) which is becoming more nurturing: the infantile form, which requires nurturing, persists to take advantage of the richer, more energetic environment. V. I. Vernadsky and Ilya Prigogine each worked out some important parts of the theory describing the emergence of order out of disorder. F. D. Peat outlines Prigogine's ideas in *Synchronicity* (Bantam Books, 1987).

Each stage of evolutionary history has its characteristic metabolic chemistry. By protecting and promoting mitochondrial respiration we are contributing to our own evolution. (More on this subject appears in "Youth, Energy and Regeneration" and "The Life of Nature.")

*Le Chatelier's principle, that a system adjusts in ways which restore a disturbed equilibrium, is behind this idea. Every part of the flow can be seen as a disequilibrium, and the complexification of the structure tends to absorb the disturbing energy.

Referring specifically to the heart muscle, Szent-Gyorgyi said "function builds structure." Generalized principles of stability will illuminate the life processes in which stimulation produces growth and adaptation. Much of the needed knowledge already exists in a fragmented way.

A proposed law of aromorphosis: that the retardation of the flow of energy in living systems tends toward a maximum; in animals, this would imply a trend toward larger-brained, longer-lived, and probably warmer animals, having a higher energy charge.

CHAPTER 5

THE LIFE OF NATURE

> *"Then tell me, what is the material world, and is it dead?" He, laughing, answer'd: "I will write a book on leaves of flowers, if you will feed me on love-thoughts & give me now and then A cup of sparkling poetic fancies; so, when I am tipsie, I'll sing to you to this soft lute, and skew you all alive The world, where every particle of dust breathes forth its joy."*
>
> W. Blake (1794)

> *"If this were an entirely scientific matter, there is little doubt from the evidence that the case for a fundamentally biological universe would be regarded as substantially proven. The reason why the scientific community passionately resists this conclusion is that biological systems are teleological, which is to say purposive. And if we admit the universe to be inhabited by a vast number of purposive components then the thought cannot be far away that perhaps the Universe itself might be purposive...."*
>
> F. Hoyle (1989)

To read either poetry or "scientific" writing, it is useful to know what was going on in the writer's life. For example, if you know that Albert Einstein's family's business was ruined by the German electric machine monopoly, his attitude toward the German-dominated physics establishment and its ideas will be seen in that context. The nature of communication and of meaning itself makes a certain consideration ("ad hominem") of the communicator's general attitudes necessary for a clear and full understanding. Einstein explicitly recognized this situation when he said that a person's life can't be separated from the person's hypothesis.

To be valid and useful, a hypothesis must coincide closely with the facts, but that is only a minimum requirement. It must also be general enough to matter, to mean something, but that still isn't enough: It shouldn't be deceptive or propagandistic, and ideally it shouldn't be misleading in any way, for example by omission and by exclusion.

The dean of the Yale school of medicine said it wasn't proper for me to think that his school's research conclusions on the safety of Premarin could be influenced by a gift of a large amount of money from Ayerst (the makers of that ultra-lucrative product), but he refused to tell me how much money was involved. Big money from every major industry has corrupted our scientific culture, and it is important that we consider the scientific alternatives.

The links between Premarin and the Theory of a Dead Universe are not obvious, unless you think about them. Our product-oriented medical culture is strengthening the idea of the pathogen-destroying magic bullet, because magic bullets can be profitably sold and administered. The idea that the patient is a self-healing, self-organizing kind of being is unpopular with the medical business, because they have no investment in the providing of healthful environments: Clean air, clean water, clean food, safe work, good housing.

The "genetic defect" theory of disease holds the promise of a gene-implant for every problem. Interestingly, the people who point out that "thousands of genetic diseases are now known" neglect to mention that the "scientific literature" supporting their claim is infinitely more "anecdotal" than the despised "anecdotal" support for the various unofficial remedies that are so offensive to the medical establishment. A doctor sees three people in a family with muscle twitches, and a Latin-named Twitchy Muscle Syndrome with dominant Mendelian inheritance enters the scientific literature. In thousands of such publications, the possibility of nutritional deficiencies and exposure to toxins isn't investigated, because there is such eagerness to recognize a new genetic defect.

There is a mystique in our scientific culture called Reductionism. It reduces the explanation for something to a description of its parts, and the ways the parts interact. A relatively small number of "elements" and "laws" are used to explain a great range of specific phenomena. My favorite formulation of the reductionist attitude is that of the famous Mr. Crick: What else is there but atoms?

The reason I call Reductionism a mystique is that its proponents feel it is so clear and obvious that they don't have to bother with philosophical complexity. They, in that sense, are the Fundamentalists of the 20th century. Although they emphasize the importance of analysis, formal reasoning, mathematics, and quantification, they are not interested in examining the philosophical basis of their methods, except occasionally to say something like "the scientific method has been proven scientifically."

Sanity itself requires that we not confuse our wishes, assumptions, methods, and ideas with the world that we are trying to understand. If our method determines our conclusions we are closer to theology than to science, and that is how many "scientists" prefer it.

An inference that something is generally true, based on some specific examples, is a matter of judgment, rather than of formal logic. Different people may draw different inferences. An intelligent hypothesis is based on a lifetime of intelligent inferences. The element of judgment in choosing a hypothesis leads some people to suggest that there is an arbitrariness or even randomness to the choice of hypotheses. If we acknowledge clearly that our choice of hypothesis is a matter of our best judgment, there is no problem, because the virtue of science is in putting hypotheses to various tests, which will eventually eliminate the most mistaken ones. When we make specific formal deductions from a hypothesis we reveal its implications and look for ways to test these implications.

It is in the nature of formal reasoning that we can deduce specific details from something more general, but that we can't formally deduce something more general from specific things. This is where reductionism uses its method to impose a conclusion about the nature of the world.

The reductionism which says that physics can explain everything, because everything is composed of "these basic particles, according to fundamental laws of interaction," forces itself into asserting that the many forms of life can be deduced from the general laws of physics, and that the varied and specific forms of organisms represent the interaction through time of chance events (in the genes) with the determining forces of the environment. The mode of explanation leads to the content of the science. The nature of formal logic, proceeding from the parts, requires that the whole be explainable in terms of its parts. If the trick that explains the whole can be found in the nature of the parts, there is no problem, but problems arise when the reductionist claims that we already have the kind of knowledge of the parts that can adequately explain the whole. In this sense, biology is seen as less general than physics, and physics contains an adequate explanation for biology.

A story from a 1915 physics conference in Berlin will illustrate how the desire to explain things in terms of the parts can create problems even within physics itself (and its traditional subject matter). Michael Polanyi, a Hungarian, found that there was a simple formula that accurately described the adsorption of gas molecules on charcoal, at different pressures, in terms of an "adsorption potential" extending through space, which allows various layers of gas molecules to condense on top of previously adsorbed layers, as the pressure is increased.

The famous German physicists at the meeting pointed out that their modern knowledge of the electrical nature of matter made it evident that the first layer of molecules would neutralize the adsorbing force of the charcoal, and that it was only because Polanyi was working in a backward country that he thought an adsorption "potential" could somehow reach out into space through a layer of molecules to pull in more molecules.

Around 1930, Langmuir proposed his relatively complex multilayer adsorption theory, but by then Polanyi's more elegant adsorption isotherm was remembered only as the embarrassing "mistake" of a young man from a backward country. (We should also remember that the "modern" notion of the electrical nature of matter formed the very basis for such work as the theory of the photoelectric effect, which had such importance in the development of the quantum theory. If atoms aren't the discrete little packets they were thought to be around the turn of the century, then the need for discrete little packets of energy isn't as evident as it seemed then. Polanyi's other projects turned up other aspects of matter that weren't acceptable to the physics of that time.)

If the world's best physicists and chemists used a reductionist approach to block the development of an important branch of physics, it is clear how the approach can obstruct the development of ideas in more complex fields, such as biology and cosmology. If the Germans had taken Polanyi's adsorption work more seriously, they would have had to revise their rigid and mistaken ideas about "the electrical nature of matter." A higher level of organization of substance (as in Polanyi's experiment) can, if insightfully described, provide new knowledge about the nature of the "parts" making up the lower level. This might be called "reducing up."

Since there is nothing known about atoms that could cause them to arrange themselves into the shapes of the various known organisms, absolute chance is said to be responsible for variations of form, and the surrounding conditions are said to allow certain ones to persist and multiply.

Once we accept that knowledge is tentative, and that we are probably going to improve our knowledge in important ways when we learn more about the world, we are less likely to reject new information that conflicts with our present ideas. The attitude of expectancy will allow us to apply insights gained at one level of generality to other levels. No particular kind of knowledge will have such authority that it will automatically exclude certain possibilities in another field of knowledge.

For example, either Polanyi's adsorption measurements or the known structures of organisms or biological systems could be seen as an opportunity to ask "what processes could create structures of this sort, or are there rules of stability which can clearly define the conditions under which such a structure can exist?"

Rules of structure-making are likely to be very complex, but rules of stability are relatively simple. Polanyi's adsorption isotherm is a very simple rule of stability, which describes how gas molecules arrange themselves on a surface, and it makes a general contribution to our understanding of the "condensed state of matter." The factors which govern mammalian fertility can be seen as part of a potential rule of stability, describing the conditions under which new mammals can come into existence; as it turns out, the factors which govern fertility also have an important influence on the subsequent development of the organism, its health, intelligence, and longevity--the way in which its very structure comes into being.

Rules of stability, with greater generality than the basic laws of physics and chemistry, might be discovered which specify the nature of the substance which is likely to exist in various situations. At present, chemical engineers have some fragmentary knowledge of this sort. The attitude of reductionism has certainly kept many things from being discovered, even in chemistry and simple biochemistry. For example, people used to say that, although you can cook a soy bean or a piece of meat into its component amino acids, you could never cook amino acids into a protein. Sidney Fox did just that, simply by omitting the water in the recipe. The reductionists then said that the polymers Fox made would be random, and without any catalytic (enzyme-like) properties, but in fact their structure was non-random, and they did have catalytic activity.

Another example: Some biochemists thought it was foolish to suggest that dilution might obscure important principles of enzyme behavior, when cells are dissolved in water in a test tube; but a brighter man down the hall thought about it for a while, and demonstrated that some of the most famous soluble test tube enzymes don't react randomly with molecules in solution in the cell, but pass the molecule from enzyme to enzyme in an organized way.

In these examples, the ignorant opinions were held by most chemists through most of the history of protein chemistry, because it was required by the reductionist belief that the "parts" explain the whole. The belief that Fox's polymers of amino acids would be "random" and inert derived from the belief that forms are generated randomly, through a long history, and that the properties of life are simply the arbitrary result of what happened to be preserved by the natural selection of the environment.

The existence of various chemical substances in interstellar space, or the presence of "biological" molecules in petroleum, or of subatomic particles in cosmic rays or radioactive minerals, can be understood in terms of various rules of stability. Plant material and human substance might best be understood in terms of such rules of stability. That would put all substances on the same "ontological level," though the more complex substances would have more specialized requirements for existence.

The "requirements for existence," seen in this way, form a direct link between practical matters, such as health and human potentiality, and seemingly remote matters, such as cosmology.

To discover such "rules of stability" for complex structures requires that we imaginatively extend the rules or laws that we now recognize. Such an imaginative extension, in searching for adequate hypotheses, can be seen as an expansion of physics, or as a "reducing up" of physics to biology.

The insistence that formal deductive reasoning should form the link between our present physical ideas, and our description of biology, is what forces the reductionists to assert that our existence is the result of random mutations plus natural selection. My preference for the use of imagination to extend physical ideas, rather than logical deduction to "apply" those ideas, comes simply from my feeling that we don't yet know everything, and that in searching for new knowledge we should avoid routes that lead us to believe nothing is there even before we have looked.

NOTES

In directing our attention toward "rules of stability," I think the following are some of the processes and principles we should consider:

1. Systems in which energy (guiding or driving energy) is added and given off at every point, systems in which no part is in equilibrium. This is how the world is, but it is usually omitted from "scientific" descriptions, because of its complexity.

2. Asymmetry as a general aspect of interaction.

3. Hysteresis, or system memory. This is one aspect of the reality of time.

4. Long range order. This simple idea is often neglected. It is implicit in hysteresis.

5. Oscillating reactions, as a paradigm for various physical processes, in which energy differences create structure in both space and time.

6. The processes of resonance as they contribute to stability of structures, and as involved in long-range order, hysteresis, oscillating reactions, asymmetry, etc.

7. All stability is metastability.

One of my professors, Francis Reithel, pointed out the fundamental importance of Le Chatelier's principle for biochemistry. Another professor, Sidney Bernhard, fundamentally changed and expanded our understanding of biochemistry by showing that glycolytic enzymes operate in a tightly organized system. Human health and the health of the planet are deteriorating in spite of the grandiose claims of the reductionists. Satisfactory solutions to complex problems will be possible only if we directly confront the complexity. Vernadsky concretized Le Chatelier's equilibrium principle in relation to biogeochemistry, and similar methods can probably be applied to many other complex systems. The number of elements and the organizational diversity of elements in the system will have to be taken into account in concretizing and applying the ideas.

CHAPTER 6

THE EX-RAINFORESTS OF THE PACIFIC NORTHWEST

In a recent television piece, Stub Stewart, one of our local timber barons, said that when he looks at an ancient forest he sees money; an average fir tree represents more than $2000, he said. (In Ernst Schachtel's analysis of perception types, this is the lowest, undeveloped or degraded type of perception, based on self-centered utility, rather than on grasping the actual nature of the thing.) Considering that this unintelligent sort of perception is typical of people who have power, I want to point out some of the neglected money related effects of mature forests of giant trees.

These forests, which the elderly timber baron described as rest homes which should be removed and replaced with kindergartenly tree farms, are often described as being biologically static systems which, for example, make no contribution toward absorbing the carbon dioxide produced by industry. (The people who use this argument would have a point, as far as the carbon dioxide question goes, if they proposed to plant new forests in shopping centers and other places that don't have trees now. But replacing a big forest by a little forest adds carbon dioxide to the environment. I don't consider the "greenhouse effect" to be a problem, but I mention it as an example of how even their own arguments weigh against their intentions.)

The fact that I think is crucial for the world's future is that rainforests energetically are comparable to volcanos,[1] except that their energy is primarily biological, and that their effects contribute massively to the stability of the biosphere. Although net growth per year (in the sense of added biomass) is greater in a tree farm, the internal flow of energy is greater in an ancient forest than anywhere else on earth, and it is this high rate of "metabolism" which gives the great forests their powerful influence on the rest of the world. Bioenergetically, they can be compared to the brain of an animal: intense, stable, and essential to the existence of the whole.[2,3]

If we imagine an ideal green slime resting on an ideal brown goo (in some future time), we could suppose that the sun's energy was being absorbed and used in photosynthesis with some efficiency at any moment. The difference between this simple system and the rainforest has to do with the complexity of life's interaction with the environment: the great age of the forest reveals the stability of its integration with the world's weather system.

Metabolic processes occurring at night influence daytime processes; winter events influence summer events; soil events influence air events. Time is bound into the system. The flow of energy is dominated by nearly horizontal switchbacks, rather than by uncontrolled vertical surges. Space is bound into the system, for example by the mysteriously powerful pumping systems involved in transpiration. (The secretion of droplets at the ends of leaf veins under certain conditions shows that osmosis and evaporation are not adequate to explain the movement of water; the attitude of our science establishment to this fundamental problem offers great opportunities for sociological investigation.)

A century ago, Darwin showed that the folding movements of many plants at sunset have the function of preventing radiant heat loss; that is, the leaves arrange themselves so that they radiate their metabolically derived energy toward other leaves, rather than toward the dark, energy-absorbing sky. He showed that the angle of the leaf at night could make the difference between life and death.

Water is a very good absorber of heat (infrared radiation). For example, a spectrum of the energy radiated from a person's hand reveals a deviation from the expected curve, as a result of infrared absorption by water. The mass of water inside a tree is therefore metabolically relevant, even though most of the metabolism of a tree is near its surface. The height and density of the forest will also affect its thermal and metabolic stability. In humans, about 75% of our metabolic energy is lost as radiant heat. In a cold environment, the energy lost as radiant heat is important to any organism. The radiant interaction of organisms in a rainforest is immense. It is another factor which distinguishes forests from tree farms.

The Food and Agriculture Organization of the U.N. defines forests as "lands bearing vegetative associations dominated by trees of any size, exploited or not, capable of producing wood or of exerting an influence on the local climate or on the water regime." Everyone who listens to weather reports on television and radio should be aware that the forests, and their destruction, are affecting their weather. Farmers and the owners of ski resorts are aware that drouths affect their income, so they shouldn't have too much trouble grasping the implications of deforestation. The loggers talk about the importance of their jobs, relative to the importance of the environment; drouths and floods, caused by deforestation, have obvious economic consequences, which must be taken into account.

Ivory dealers say "but I have to make a living" even if elephants become extinct. Drift net fishermen say "I have to make a living," even if entire populations of porpoises are destroyed. Every region and industry seems to have people who feel that their present jobs are more important than the very ecosystem which is the foundation for their present "harvesting" job. This pathological sort of perception resembles the type of autism in which the victim devours part of his own body, unless restrained. Just as Gorbachev said that the survival of humanity transcends the class struggle, protection of the environment must transcend the witless arguments of those who insist that they must be allowed to consume these resources now, before it's too late for them to enjoy this one last harvest.

In Alaska, gigantic Sitka spruce are being turned into wood pulp to make into rayon shirts. In Oregon and Washington, many ancient trees are made into pulp for newsprint and wrapping paper. Redwood is so resistant to decay that houses built from it should last forever, but most of the houses built from the extinct redwood forests have been bulldozed into splinters and hauled to landfills, as the cities grow and change. One California company which had been slowly cutting its large timber holdings was recently forced to begin cutting trees as fast as possible, because of debt created when it was acquired by junk-bond billionaire Hurwitz. The fortunes created by this destruction of wealth are stimulating the concentration of wealth and power, but do not seem to be leading to the production of new wealth. The condition of towns and regions which have depended on the timber industry speaks eloquently of what the timber baron's "wealth" amounts to in human terms.

The alternative sources of cellulose for rayon and paper are numerous. Besides cotton, linen, and hemp, various types of paper have been made from straw, esparto grass, jute, sisal, cane, and bamboo. The use of wood for building houses is largely a cultural matter, since there are many better and cheaper materials. For example, I lived in a two-storey adobe house built around 1450, which had survived several major earthquakes without damage. It had 12 ft. ceilings, and didn't need air conditioning even in hot weather. When the drainage is good, these mud buildings are not damaged even by torrential rains. My father built an adobe house in the 1930's which is still in good condition. Even the shingles were made of clay, mixed with an emulsion of tar for waterproofing. These houses can never be destroyed by fire, and they don't have the toxic fumes often produced by chemically treated wood products.

Even northern forests can provide rich harvests of products other than wood, without causing ecological damage. Resins, drugs, hormones, and food materials are produced in abundance, though the moronic practices of clear-cutting and slash-burning have eliminated some of the potential forest crops. According to the calculations of biologist Curt Mitchell, the forests of Oregon can produce at least $4,000 worth of materials per acre per year other than trees.

About 96% of the redwoods which used to extend from San Francisco to southern Oregon are gone. Only about 2 square miles of redwoods remain in Oregon, and many of these are scheduled to be sold and logged this year. The redwoods illustrate some of the interactions between trees and climate. The coast redwood (Sequoia sempervirens) can reach a height of more than 350 feet. It requires areas which provide summer fog as well as winter rain. In some areas, it is estimated that 50% of the annual precipitation is in the form of fog caught on their leaves, so it is obvious that their great height makes a major contribution to the water-economy of the region.

The Sierra redwood (Sequoiadendron giganteum) grows at a higher elevation, on the western slopes of the Sierra Nevada Mountains of California. Their massive trunks apparently represent an adaptation to the absence of the summer fog--their wood is about 50% water, so the thick trunks function as storage tanks to sustain activity through the dry seasons (Figure 6.1).

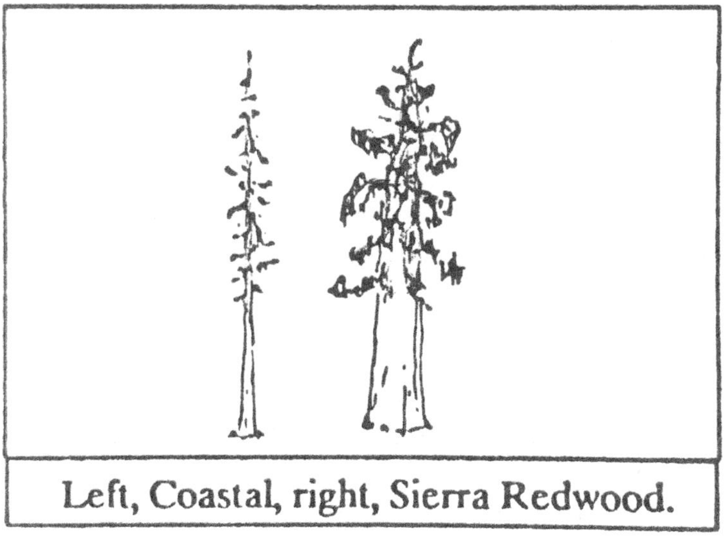

Figure 6.1 Adaptive change of shape as illustrated by trunk size of Coastal Redwoods (depicted on the left) to Sierra Redwoods found at higher elevations (depicted on the right).

Disregarding the soil's water-retaining capacity in an ancient forest, just the storage of water in the trees themselves can amount to thousands of tons per acre; such quantities of water have significance even geologically. According to A. V. Lapo, in *Traces of Bygone Biospheres*, "the content of water in the tissues of living organisms is approximately five times greater than in all the rivers of the globe." "Half the water...in the roots of plants is renewed in several minutes. As a result, the water cycle on land is determined almost exclusively by the transpiration of the plants."

The humidity of the air passing through a tree farm is much lower than that passing through a real forest. Locally, this determines what can live and thrive. On a larger scale, differences in temperature and humidity affect wind flow. I think this is why the humid ocean winds are increasingly being diverted toward British Columbia, and away from Oregon and California, where they used to provide dependable winter precipitation. (Since ocean currents are guided by winds, the disruption of weather patterns is likely to become self-stimulating.)

2500 years ago, Theophrastus recorded cases in which a region's climate was permanently altered by deforestation and swamp draining, with the result that rich agricultural productivity was lost forever. The difference between the small scale of the ancient Greek ecological damage and the present scale of deforestation suggests the scale of the damage which is being done to modern agriculture.

Deforestation of the Pacific Coast is moving from south to north. The areas that are immediately "downwind" from the treeless coast are the places that will feel the earliest effects on their weather. California and Nevada are now facing their fourth consecutive drouth, and have just experienced the driest December in history. Farmers have been warned to expect a 20% cut in water deliveries this summer. The decreased production of food is likely to cost the region several billion dollars this year, and there is no reason to believe that the old weather patterns will return, without the forests.

Since deforestation causes the soil to be washed away, muddy water causes massive destruction of fish. Reservoirs behind dams have filled rapidly with mud, shrinking their storage capacity, and their ability to control floods and to provide water for irrigation.

In the 1920's, an economist, S. G. Strumilin, showed that money spent on education gave a 27-fold return; more recently, J. K. Galbraith said that the return is about 100 times the investment, and is greater than the return on investments in railroads, dams, or factories. Buckminster Fuller talked about doing more with less; this is the meaning of applying science to economic production maximizing effects while minimizing the means.

To have an economy that develops qualitatively, rather than quantitatively, means that the extractive industries will decrease, while the complexity, intensity and appropriateness of use of the materials will increase steadily. Investment in humanity means increasing productivity and enjoyment, by improving human health, intelligence, skill and knowledge.

While the economists showed the value of investing in traditional schools, the returns would be greater, and the costs smaller, if we began to apply thought and knowledge to understanding the learning process better. The useful part of education is that which stimulates the imagination. The useful part of knowledge is that which expands human possibilities. If the biosphere continues to be degraded, human possibilities will be stunted, at best.

I think we have to start seeing jobs in resource-extraction as an index of poverty, to the extent that they degrade the biosphere. A definition of wealth must consider both quantity and quality of production. Environmentalists and ecologists have to do some intelligent thinking about economics, so they can help the people in power to think more realistically.

The greatest fraud now current in the mass media is the idea that we must choose between the health of our environment and the vitality of the economy or our standard of living. Just as competition from Japan and Germany is causing big corporations to move quickly to get out of the weapons business and to reinvest in the consumer economy, the present anti-environmental direction of our economy depends on relatively few decisions made by big corporations (and supported by their agents in government). Since they can get away with making stupid decisions, they make them.

The stylish science-talk about global warming doesn't really take very much into account, beyond a few cultural stereotypes. The most important context to take into account is the importance of moving toward an economy which develops qualitatively rather than quantitatively. There is no foreseeable limit to the qualitative development of the economy. If we can shift from an extractive-degradative economy to one which develops by investing in education, science, culture, and human well-being, we will have a future in which to discover whether the biosphere-noosphere has any limits to its developmental potential.

NOTES

1. In the sense that the heat of volcanos can "drive" changes in ocean currents: winds, "driven" by forests, are the other major influence on the patterns of ocean currents, and thus on the distribution of solar energy around the planet.

2. Following the brain analogy, a system of tree farms is like the anencephalic monster; it can grow if fed, but can never contribute to the world's advance.

3. Rainforests illustrate the first and second principles of biogeochemistry; that biogenic migration of chemical elements tends to the maximum, and that the development of stable life-forms serves this tendency.

PART II

ENERGY PROBLEMS

"Energy is the only life, and is from the Body. . . .
Energy is eternal delight."

W. Blake

An emphasis on energy is rapidly gaining acceptance in biology, if not in medicine. Just twenty years ago, many biologists denied that Atkinson's idea of the cellular "energy charge" had meaning or relevance for living systems, but the idea is now widely recognized, and most biologists would probably admit that it has at least some relevance for some situations. In this section, I discuss some of the practical consequences and applications of the idea that we should focus on the role of energy in biological adaptation. Energy, in an important sense, is prior to information, especially in the context of biology.

CHAPTER 7

A UNIFYING PRINCIPLE

For many cells, the threshold for excitation (and response) is governed by the energy charge of the cell. This is clearest in the case of brain cells: seizures can be produced by either hypoxia or hypoglycemia, and in the low energy state of hypothyroidism, deep normal sleep is seldom possible. Muscle cells, secretory cells, and immune cells behave similarly. I believe we are at the beginning of a therapeutic revolution based on this, and other closely related principles. It is easiest to understand these relationships if we use a graph, with "energy charge" on one axis, and with "intensity of stimulation" on the other. This graph represents the relationship of an organism to its environment, or of a cell or tissue to its (internal) environment (Figure 7.1).

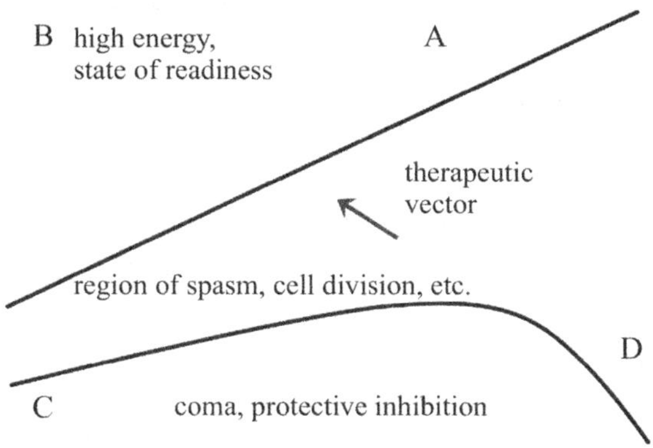

Figure 7.1 Relationship and response of an organism to its environment.

The regions labeled with A, B, C, and D can be thought of, for the sake of illustration, as the following functional states of the organism:

A -- Adventure
B -- A baby sleeping
C -- Cautious conserving of energy
D -- Stressed to death

The ideal therapy is one which restores the cellular energy thoroughly, so the organism regains its full capacity to adapt, and seeks appropriate stimulation. Good nutrition and sleep are sufficient therapy for a healthy young organism. For old or sick organisms, artificial sleep and special nutrition can be useful. The use of sedatives to produce an "artificial coma" is an old technique which is now being used more often to treat drowning victims and other brain-injured patients.

Sedation or anesthesia decreases the use of energy, making it possible to recover a store of energy, and the efficient, relaxed state of readiness that goes with an adequate reserve of energy. Thyroid and other hormones can increase the rate at which energy is produced, but some of the steroid hormones also have the sedative and anesthetic actions which provide the other aspect of recovery: while production of energy increases, wasteful use decreases.

The idea of the "therapeutic vector" means that any treatment should combine preservation of energy with restoration of energy production. Shock, which is a state of suppressed energy production, can be used to illustrate this approach.

Several times in the last 50 years great advances were made in understanding the nature of shock, but they tend to be ignored or forgotten. I think this is because our medical culture has lacked a unifying principle which would integrate physiology and biochemistry into a functional whole.

Under extreme conditions, the state of shock can develop. Blood circulation becomes so inefficient that the organism often dies from the shock state itself. It has been traditional to administer adrenalin to shock patients to raise their blood pressure, but adrenalin is often already present at a high level, and more can simply exacerbate the problem. This is because the arterioles contract in shock, and under the influence of adrenalin (and cortisone).

Early in Hans Selye's study of the steroids, he noticed that estrogen treatment causes the same changes which occur in the shock phase of the stress reaction. Many years later I noticed that oxygen deprivation causes the same biochemical changes that occur with an estrogen excess, and with old age, and I proposed that estrogen acts by interfering with the use of oxygen. In the mid-1940s a group of researchers, believing that energy depletion was the basic cause of shock, demonstrated that intravenous injection of the "energy molecule" ATP caused animals to recover which otherwise would have died from shock.

ATP isn't the sort of chemical that is likely to interest the pharmaceutical industry, and although other people have demonstrated that it has therapeutic value in a variety of diseases, the dominant attitude has been to scoff that it is "only a vasodilator." But the tone of smooth muscle in blood vessels is largely a function of the energy charge of the muscle cell. Thus, intravenous glucose can restore the balance between the muscle tone of arteries and veins, and it has been used many times to bring patients out of shock. If a physician is focusing on his patient's extremely low blood pressure, a vasodilator is the last thing he would consider, and salt water is often preferred in place of glucose, but restoring energy should be the basic aim.

Keeping in mind the relationships between excitation and inhibition discussed above (Figure 7.1), we can visualize how the energy supply to vascular smooth muscle normally functions as part of a self-regulating system, and how a failure of energy can produce the self-defeating shock state (Figure 7.2).

In the blood vessels at the left side of Figure 7.2, the arterioles are in the high energy resting state, while the veins are in the middle, active zone. On the right, the arterioles are receiving "depleted" blood, lacking oxygen or sugar for example, and so are now in the middle energy, active state. Passing through the capillaries, most of the blood's remaining oxygen and nutrients are

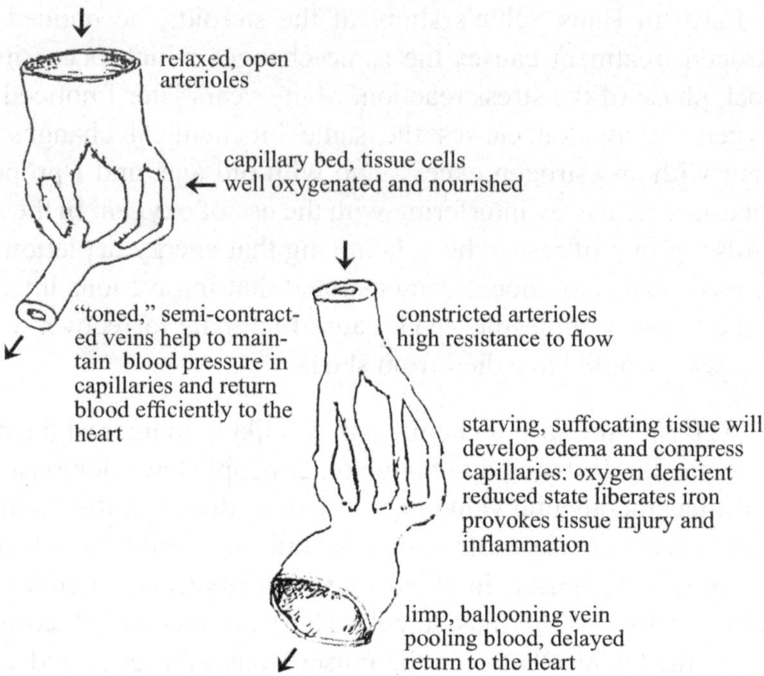

Figure 7.2 Depiction of "toned" vascular smooth muscle with adequate energy supply (upper left), versus constricted state (lower right) resulting from low energy supply.

extracted by the tissue cells, leaving only minimal energy supplies for the veins, which are therefore in the lowest energy state, "protective inhibition," in which there is no muscle tone.

Besides helping to explain the shock state itself, this relationship of vascular tone to cellular energy helps to explain the circulatory distress which is associated with fatigue, old age, hypothyroidism, and estrogen excess.

Bacterial endotoxin causes some of the same effects as adrenalin. When stress reduces circulation to the bowel, causing injury to the barrier function of the intestinal cells, endotoxin can enter the blood, contributing to a shock state, with further impairment of circulation. In old age and in "winter sickness," something like a chronic borderline state of shock can develop. Intravenous glucose has been used successfully to bring patients out of septic shock. The tonic effects of intravenous local anesthetics are, I think, largely the result of their ability to open the arterioles. Magnesium and vitamin A also have some ability to normalize blood vessel tone, and can help to maintain the barrier function of the bowel.

Whether a tissue is hyperactive or underactive, it is possible that a systemic energy deficit is responsible. Symptomatic relief is rational if it is consistent with restoration of energy.

Glucose is often thought of as the most direct source of energy, but other substances are apparently used even more easily. "Ketones" (for example, alpha-keto or hydroxy-butyrate) are used more easily, at least in some circumstances. Short and medium chain fatty acids are used more easily than glucose, and it is apparently this fact which accounts for their presence in milk. Their effects on cells--induction of hormone receptors and other specialized cell functions, suppression of stress-induced enzymes, stimulation of energy production in fat cells, inhibition of cancer cell division and viral expression, etc.--are what we would expect of an ideal energy source. Unfortunately, commercial milk animals are fed large amounts of grain, the oils of which act in opposition to the short and medium chain fats. Some tropical fruits and coconut oil provide some of these efficient and protective energy sources. As little as one or two teaspoonfuls of coconut oil per day appears to have a strong protective effect against obesity and cancer.

CHAPTER 8

STEROIDS

This type of molecule might be the most common carbon compound in the universe. It is made by single celled organisms, by plants, and by animals, and has many kinds of function. The steroid hormones are involved in all aspects of animal physiology, and overlap with control functions of the nervous system, peptide hormones, metabolites, prostaglandins, cyclic nucleotides, etc. Sometimes people speak of "steroids" when they mean glucocorticoids such as cortisol or a synthetic like dexamethasone, or, among athletes, when they mean anabolic steroids or synthetic androgens; and so it is common to associate "steroids" with harmful side effects. All foods contain steroids and sterols (a major type, containing an alcohol group and a side-chain) some of which are beneficial and some of which are toxic or allergenic.

In animals, cholesterol is the basic sterol molecule, which is massively converted into other substances, including the steroid hormones. Thyroid hormone and vitamin A are required for this conversion. The first step occurs in the energy-producing mitochondrion, where cholesterol loses its side-chain and is slightly oxidized, producing pregnenolone. Being less fat soluble than cholesterol, pregnenolone leaves the mitochondrion, so it can't inhibit its own synthesis. Rather, it seems to stimulate its own synthesis, though this isn't as clearly established as in the case of progesterone.

Depending on the tissue, pregnenolone will be converted by enzymes in the cytoplasm into either progesterone or DHEA (dehydroepiandrosterone). The fact that progesterone (and probably pregnenolone) stimulates its own synthesis means that taking it does not suppress the body's ability to synthesize it, as happens with cortisol. Sometimes, one dose or a few doses can restore the body's ability to produce enough of its own.

Progesterone also allows the thyroid gland to secrete its hormones, especially when the thyroid function has been inhibited by estrogen. Since the thyroid hormone is needed to produce progesterone, a supplement of either tends to normalize both thyroid and progesterone production.

Progesterone and DHEA are the precursors for the other more specialized steroid hormones, including cortisol, aldosterone (sodium-retaining hormone), estrogen, and testosterone. The formation of these other hormones is tightly regulated, so that taking the precursor will correct a deficiency of a specialized hormone, but will not create an excess. At least in the case of progesterone, an excess tends to balance or neutralize an excess of the specialized hormone, so it has been described as having anti-androgenic, anti-estrogen, anti-aldosterone, and anti-cortisol functions.

Many steroids have a protective ("catatoxic") action against a wide variety of poisons. Some of the quick effects (e.g., within 10 minutes) of progesterone and pregnenolone probably represent a catatoxic action, as well as a neutralizing or balancing of excessive estrogen or cortisol. Improved metabolic efficiency, sparing oxygen and glucose, will have a quick effect in reducing edema.

During pregnancy, very large amounts of progesterone are made. It protects and stabilizes practically all functions of both the mother and the fetus. Progesterone, glucose and the thyroid hormones powerfully influence the brain development and intelligence of the baby, probably by influencing both the number and the size of brain cells, and the quality of their functioning.

Part of progesterone's protective effect is a result of its quieting effect on cells. For example, it tends to prevent seizure activity in brain cells. During childbirth, its normal function is to act as an anesthetic. When the level of estrogen is too high, progesterone can't achieve this effect. In a non-pregnant person, it is important to determine the minimum effective dose by taking only a few drops at a time, and repeating this small dose about every 20 minutes until symptoms have been controlled. Otherwise, serious "drunkenness" can be produced, with loss of coordination, and even unconsciousness.

The only solvent for progesterone which isn't toxic and which will dissolve an effective quantity, is vitamin E. In this form, it can be absorbed through the skin or other membranes, or can be taken orally. Taken orally, it is absorbed as chylomicrons, going into the general circulation (as vitamin E does), instead of to the liver where it would be prepared for excretion. In this form, therefore, it is fully and quickly available to all tissues. It is approximately 20 times more powerful in its action than other preparations, so it is important to use it in physiological quantities, rather than in the huge doses commonly given rectally or by injection. Ten or 20 mg. is often an effective dose, though people with low thyroid or high estrogen sometimes use 50 to 100 mg. per day. In the customary 10% solution, one drop contains about 3 mg. progesterone, and 1 ml. (1/4 tsp.) contains 100 mg. The first dose should never be more than 15 mg.

Pregnenolone, taken orally, does nothing noticeable to a healthy animal or person, but if the stress-related hormones are elevated, they return to normal when pregnenolone is taken. The brain contains much more pregnenolone, DHEA, and progesterone than do other organs or the blood, and these levels decrease progressively with age. Older people are more likely to feel an effect from pregnenolone, than are young people. A tenth of a gram is a reasonable first dose, though some people seem to need as much as 1 gram per day.

CHAPTER 9

THYROID

Measuring the amount of thyroid hormone in the blood isn't a good way to evaluate adequacy of thyroid function, since the response of tissues to the hormone can be suppressed (for example, by unsaturated fats).

In the 1930's accurate diagnosis was made by evaluating a variety of indications, including basal oxygen consumption, serum cholesterol level, pulse rate, temperature, carotenemia, bowel function, and quality of hair and skin. A good estimate can be made using only the temperature and pulse rate.

Oral or armpit temperature, in the morning before getting out of bed, should be around 98° F, and it should rise to 98.6° by mid-morning. This is not valid if you sleep under an electric blanket, or if the weather is hot and humid. A person who is hypothyroid produces heat at a low rate, but doesn't lose it at the normal rate, since there is less sweating, and the skin is relatively cool. Many hypothyroid people compensate with high adrenalin production (sometimes 40 times higher than normal), and this tends to keep the skin cool, especially on the hands, feet, and nose. The high adrenalin is the consequence of low blood glucose, so a feeding of carbohydrate, such as a glass of orange juice, will sometimes lower the pulse rate momentarily. Healthy populations have an average resting pulse rate of about 85 per minute. Especially in hot weather it is useful to consider both temperature and pulse rate.

The Achilles tendon reflex is another quick way to estimate thyroid function. This reflex is used because of the insignificant weight of the toes in relation to contraction of the gastrocnemius muscle. The T (repolarization) wave on the electrocardiogram is a similar indicator of the rate of energy production. Thumping the Achilles tendon causes the muscle to contract (unless it is already in a semi-contracted state, which isn't uncommon). The contraction consumes energy, and the muscle can't relax until enough energy has been produced to restore the threshold and the readiness for a new contraction.

If energy production is efficient, relaxation is faster than the passive return motion of the foot, so the foot swings freely back to its original position, and over-shoots slightly, causing a slight swinging action. In hypothyroidism, the foot returns as if controlled by a pneumatic door-closer, and settles slowly and precisely into its relaxed position, sometimes with a hesitating, intermittent motion. This slow replenishment of energy, and slow relaxation, can cause muscles to cramp easily. The aching leg muscles of children at the end of an active day are often a sign of hypothyroidism, and sometimes the gastrocnemius muscle becomes very swollen and hypertrophied in hypothyroid children. The same process, of slow energy regeneration, can cause rhythm disturbance in the heart, and often causes insomnia and restless sleep.

The thyroid gland secretes about 3 parts of thyroxin to one part of triiodothyronine, and this allows the liver to regulate thyroid function, by converting more of the T4 to the active T3 when there is an abundance of energy. Glucose is essential for the conversion, so during fasting there is a sharp decrease in metabolic rate, and in experiments, 200 or 300 calories of carbohydrates can be added to the diet without causing fat storage.

When the liver is the main cause of hypothyroidism, your temperature (and especially the temperature of your nose, hands and feet) will fall when you are hungry, and will rise when you eat carbohydrates. If a hypothyroid person has a very slow pulse, and feels lethargic, it seems that there is little adrenalin; in this case, a feeding of carbohydrate is likely to increase both the pulse rate and the temperature, as the liver is permitted to form the active T3 hormone.

Women often have above-average thyroxin, with symptoms of hypothyroidism. This is apparently because it isn't being converted to the active form (T3). Before using a Cytomel (T3) supplement, it might be possible to solve the problem with diet alone. A piece of fruit or a glass of juice or milk between meals, and adequate animal protein (or potato protein) in the diet is sometimes enough to allow the liver to produce the hormone. If Cytomel is used, it is efficient to approximate the physiological rate of T3 formation, by nibbling one (10 or 25 mcg.) tablet during the day. When a large amount is taken at one time, the liver is likely to convert much of it to the inactive reverse-T3 form, in a normal defensive response.

Women normally have less active livers than men do. Estrogen can have a directly toxic effect on the liver, but the normal reason for the difference is probably that temperature and thyroid function strongly influence the liver, and are generally lower in women than in men. Estrogen inhibits the secretion of hormone by the thyroid gland itself, probably by inhibiting the proteolytic enzyme which dissolves the colloid. Progesterone has the opposite effect, promoting the release of the hormones from the gland. At puberty, in pregnancy, and at menopause, the thyroid gland often enlarges, probably as a result of estrogen dominance.

Thyroid function stimulates the liver to inactivate estrogen for excretion, so estrogen dominance can create a vicious circle, in which excess estrogen (or deficient progesterone) blocks thyroid secretion, causing the liver to allow estrogen to accumulate to even higher levels. Progesterone (even one dose, in some cases) can break the cycle. However, if the gland is very big, the person can experience a few months of hyperthyroidism, as the gland returns to normal. It is better to allow the enlarged gland to shrink more slowly by using a thyroid supplement. If an enlarged gland does begin to secrete too much thyroid hormone, it can be controlled with tablets of propylthiouracil, or even with raw cabbage or cabbage juice, and cysteine-rich meats, including liver.

Besides fasting, or chronic protein deficiency, the common causes of hypothyroidism are excessive stress or "aerobic" (i.e., anaerobic) exercise, and diets containing beans, lentils, nuts, unsaturated fats (including carotene), and undercooked broccoli, cauliflower, cabbage, or mustard greens. Many health conscious people become hypothyroid with a synergistic program of undercooked vegetables, legumes instead of animal proteins, oils instead of butter, carotene instead of vitamin A, and breathless exercise instead of a stimulating life.

CHAPTER 10

THE STRESS OF DARKNESS

Cortisol begins to rise as soon as there is darkness, regardless of sleep or waking. Artificial light, and its absence, clearly can determine the time at which cortisol begins to rise.

Falling glucose is the signal for cortisol to rise. Dreams are closely associated with the chemical-physiological mobilization in response to low blood glucose. (Rising adrenalin probably mediates the cortisol response to decreasing glucose.)

Sleep minimizes the increase of cortisol. The inhibitory, sleep-inducing brain chemicals, GABA and the sleep peptide, have a quieting, anti-stress effect even during waking stress.[1] Sleep (and the inhibitory chemicals) reduce energy expenditures, conserving glucose.

The mitochondria, by their ability to use oxygen, are responsible for the normal efficient use of glucose.

Beginning at sunset, darkness progressively damages mitochondria structurally and functionally. Mitochondrial damage, and the functional impairment of organs such as the heart muscle, reach a peak at dawn.[2]

During the day (in organisms which survive the night), mitochondrial structure is restored, and organ function improves, with the peak being reached at sunset. A residue of altered cell and tissue structure, and of impaired enzyme function, remains under natural conditions of day-night cycles.

In winter, there is cumulative damage to mitochondria, because of too few daylight hours to complete the rebuilding of mitochondria.[3] During the summer (in organisms that live long enough) the cumulative damage is mostly repaired.

In the dark of polar winter, ordinary artificial light provides some protection during the 16 hours when it is used, but during the dark hours when lights are off, healthy young people show the darkness-related signs that normally occur in older people, such as an accelerated pulse.[4]

Although direct sunlight damages skin, people generally are healthier in proportion to their exposure to sunlight.

Ultraviolet and blue light produce free radicals in living material, but they penetrate only a few millimeters into tissue. Yellow, orange, red, and infrared rays penetrate much more deeply.

Small organisms are able to set their biorhythms by a very weak light. Only bright lights can alter human rhythms. For example, the cycle of body temperature can be changed, or the rhythmic secretion of melatonin can be suppressed by lights stronger than 2500 lux.[5]

Three common biological pigments are able to absorb the penetrating red light: Biliverdin, blue copper proteins, and melanin.

The presence of melanin in nerves suggests that it should be considered as a candidate for the pigment which allows light entrainment of rhythmic melatonin secretion.

The presence of copper proteins in every cell, and in the mitochondria, suggests that they should be candidates for the red light receptor which promotes respiratory function and restoration of the mitochondria.

The presence of copper in the crucial respiratory enzyme, cytochrome oxidase, is of the greatest interest.

Both copper and light appear to be involved in other aspects of respiration, including cyanide detoxification,[6] and possibly even in activation of the thyroid hormone.[7]

Besides darkness, hypoxia and overload stress (which is relative hypoxia, in a sense) are the main things which damage mitochondria. Inhibitory analogs of GABA appear to be protective in all types of stress. The anesthetic steroids, such as progesterone, pregnenolone (and possibly DHEA), also protect against injury by stress, and probably work by some of the same energy-sparing mechanisms.

Anti-dark-stress substances would tend to minimize nocturnal hypoglycemia by various mechanisms, including the inhibition of insulin secretion. Acetylcholine and estrogen stimulate insulin secretion, so cholinolytics (e.g., atropine) and anti-estrogens (e.g., progesterone and DHEA) would tend to inhibit it, and GABA helps by causing insulin uptake by red blood cells.[8]

Both living at a high elevation and having high thyroid function increase the quantity of mitochondria in cells, and this increase appears to be fairly permanent. Pretreatment with thyroid hormone or with a time at high altitude increases the ability to resist stress.

Living at high elevation or having high thyroid function also increases the vascularity of tissues, that is, the number of capillaries per unit of tissue, and increases the quantity of blood in the tissue, protecting against hypoxia.

Avoiding excessive intake of unsaturated fatty acids and iron will decrease the damage caused by unavoidable stress (including darkness).

Analogs of the inhibitory nerve transmitter, GABA, protect against damage to the mitochondria in the natural way that sleep does, minimizing the injury caused by cortisol. Milk and various fruits contain close analogs of GABA, some of which specifically block formation of the catabolic stress enzymes.

To the extent that nocturnal stress resembles shock, taking

some extra salt at bedtime is logical. It lowers adrenalin, and tends to increase blood volume. In hypoxic stress, tissue retains water, and increasing the sodium content of the blood opposes that tendency. It is often a remedy for insomnia.

Penetrating red light is possibly the fundamental anti-stress factor for all organisms. The chronic deficiency of such light is, I think, the best explanation for the deterioration which occurs with aging. Enzyme changes, free radical changes, structural and respiratory changes are all involved as consequences of darkness stress.

Just as there are multiple hormonal and nervous systems which ensure the reliability of our physiology, we should expect red light to have more than one beneficial function. For example, it could quench excited electrons, desorb inhibitory toxins, maintain enzyme function, mobilize stored copper, and regulate the conductivity of subcellular structures. Red light, applied within the first several minutes following x-ray or gamma irradiation, protects organisms (from bacteria to mammals) from radiation injury.[9] The activation of energy-dependent repair processes is clearly involved, but the exact physical and chemical mechanisms still haven't been determined.

During the winter when there isn't enough sunlight, it is helpful to use several hundred watts of incandescent lights. If bulbs with internal reflectors are used, 500 or 600 watts can be bright enough to have a protective biological effect.

NOTES

1. Sukhikh, G. T., Meerson, F. Z. & Mikhaleva, I. I. (1985). Stress reduction of the resistance of the organism to tumor growth and prevention of this phenomenon by central inhibitory metabolites. *Bulletin of Experimental Biology and Medicine*, 595-597.

2. Frolov, V. A., Pukhlyanko, V. P., & Kanskaya, T. A. (1985). Changes in left ventricular mitochondria in intact rabbits during the 24 hour period, *Bull. Exp. Biol. & Med.*, 353-356.

3. Frolov, V. A. (1984). Seasonal structural and functional changes in the rabbit heart. *Bull. Exp. Biol. & Med. 97*(4), 420-423.

4. Maksimov, A. L., & Chernook, T. B. (1986). Biorhythmic aspects of intercontinental Antarctic adaptation. *Izvestiya Akademii Nauk Kirgirskoy SSR, No. 2,* (March-April), 35-37.

5. Honma, K., et al. (1987). Entrainment of human circadian rhythms by artificial bright light cycles. *Experientia, 43*(5), 572-574.

6. Davies, I. J. T. (1972). *The Clinical Significance of the Essential Biological Metals.* C.C. Thomas Publ., 112.

7. Hemphill, D. D. (Ed.). (1979). *Trace Substances in Environmental Health.* Univ. of Missouri, 360-367.

   Shkolnik, M.Ya. (1984). Trace Elements in Plants. *Elsevier*, 177.

8. Sandulyak, L. I. (1987) Erythrocytes as depots and carriers of insulin. *Uspekhi Sovremennoy Biologii, 103*(2), March-April, 207-216.

9. Bulyakova, N. V. & Popova, M. F. (1987). Stimulation of post-traumatic regeneration of skeletal muscles of old rats after x-ray irradiation. *Bull. Exp. Biol. & Med., 103*(4), 548-550.

CHAPTER 11

PREGNENOLONE

My interest in pregnenolone was the result of an accident. I had been experimenting with the solubility of various steroids. I found that pregnenolone isn't very soluble in vitamin E, and I saved the bottle containing vitamin E and pregnenolone to use as a vitamin E supplement. I had another bottle of pure vitamin E.

In 1983, the ocean current called El Niño caused Oregon's weather to stay cloudy and rainy through the summer. I didn't have my usual summer recovery from "winter sickness," and by November I was becoming very sick. During a two week trip I felt perfectly well, but except for that short time I kept getting sicker and sicker (with inflammation of arteries, dental abscesses, asthma, migraines, and colitis). Lying in bed one day trying to understand what had happened during that two week trip, I remembered that I had used the vitamin E containing a residue of pregnenolone on the trip, but used pure vitamin E after I got home. I immediately got out of bed and took a pinch of pregnenolone. In about an hour, all of my symptoms had stopped, and for over a week I was healthy and energetic. But then I got sick again, and was so distracted by the pain that it was several days before I remembered the pregnenolone. Another pinch had the same dramatic sudden effect. Since then I have taken some almost every day. While I was sick I had pictures made for a passport, but they looked so odd that I got several more. The best of those, the least haggard looking, is reproduced here.

Pregnenolone is known to be produced (in the mitochondria) from cholesterol. This is the first stage in the production of all of the steroid hormones. If pregnenolone synthesis is insufficient, supplementary pregnenolone would help to maintain an optimum

Figure 11.1 Before and after pregnenolone.

BEFORE: Years of declining health with "winter sickness." Photograph July 1983. Note the drooping skin. Eyelids were covered by the sagging skin.

AFTER: About ten weeks using 50 to 100 mg. per day. Photograph mid-September, 1984. I suspect that connective tissue cells contract almost like smooth muscle under the effects of pregnenolone; there seems to be turgor without edema.

level of the various other steroids. Aging, stress, depression, hypothyroidism, and exposure to toxins are conditions in which synthesis of pregnenolone might be inadequate.

In 1944 factory workers were given daily doses of pregnenolone and, on alternate weeks, a placebo. During the pregnenolone weeks, their piece-work output was significantly higher. Healthy young people who are not under stress do not feel anything from taking even a very large dose (e.g., one gram), and no changes have been detected in blood glucose, proteins, amino acids, or salts. The urine apparently contains more conjugated steroids (indicated by an increased tendency to foam) after taking a large dose of pregnenolone; I think light is the main factor normally causing changes in the conjugated steroid content of urine.

The skin and the brain are important sites of steroid synthesis, and they are usually ignored by endocrinologists investigating steroid hormones. Papa and Kligman (1965) showed that adding pregnenolone or progesterone topically to skin would restore youthful properties to aged skin. I think DHEA gives similar results, and (like progesterone) is more easily absorbed than pregnenolone. Both progesterone and pregnenolone seem to improve the efficiency of blood circulation. I have seen grey-skinned depressed people turn pink and start smiling shortly after taking pregnenolone.

About 1950, pregnenolone was tested on several people with exophthalmia from Graves' disease, and it was reported that their eyes quickly receded toward a more normal position in their sockets. Having read this, I mentioned it to a desperate woman with horribly bulging eyes. Her left eyeball was about 50% exposed, and the other one was less extreme, but still seriously protuberant. She took some pregnenolone as we talked, and within an hour it seemed to me that her eyes were receding. An hour after she took the pregnenolone, she said her eyes were feeling much better. I left her city early the next morning, but she telephoned me the next day and claimed that her eyes were completely normal. Without seeing her, I don't know exactly what she meant by "normal." This happened over two years ago, and I hope to visit her city soon to see whether her eyes really became, and stayed, normal.

For many years it has been known that steroids, including cholesterol, have an antitoxic effect. The cytochrome P-450 family of enzymes are an important factor in our resistance to toxins. Moderate amounts of cortisol promote this system, but larger amounts degrade it. Pregnenolone doesn't affect the rate of synthesis of these enzymes, but it stabilizes them against the normal proteolytic enzymes, increasing their activity. I believe this stabilizing action is a general feature of these steroids, explaining other antitoxic effects such as blocking hemolysis, and probably many features of growth and differentiation, including control of cell division and prevention of atrophy.

NOTE

After using pregnenolone for several years, I noticed that the slack skin and other signs of aging were coming back. That spring I increased my dose of pregnenolone, and did other things to promote my own pregnenolone synthesis, such as using larger amounts of coconut oil and getting a little more exposure to incandescent light. The effects of the supplements are more noticeable in the summer, since that is when the body's repair processes are greatest. Avoiding the destructive effects of northern winters (without migrating to the south) would seem to require artificial lights as bright as summer daylight.

CHAPTER 12

RESTORING HAIR COLOR

Largely because of Albert Szent-Gyorgyi's explorations in biological pigments as factors in cell regulation and immunity, I have paid special attention to several pigments, such as melanin, the black or brown pigment of hair and skin, of injured potatoes, and of certain fungi, when damaged. When I lived in Montana, I knew many people whose hair became grey in their late 20s; I wondered whether it was caused by low thyroid, by an odd mineral balance, by frequent chilling of the skin and scalp, or maybe just by their lack of fresh fruit and vegetables. In Mexico, people in warm, humid tropical areas seldom lost their pigment. When I heard Carl Pfeiffer speak, I noticed that both his skin and his hair were free of pigment, though he was not an albino, and I wondered whether his various mineral nutritional supplements had contributed to his fading. (He took sulfur every day, and avoided copper. He also advocated the use of molybdenum as a nutritional supplement.)

Until I was 4 or 5, my hair was white, and I took it as a humiliating stigma of sexual immaturity; in fact, everyone tends to darken as they grow up, especially around puberty. Many people have wondered what function the melanin in frogs' eggs might have, and a few years ago someone demonstrated that it is what binds progesterone in the egg, and suggested that it might have a similar function in other animals and other cells.

A Japanese researcher found that each hair color is associated with a certain pattern of several trace minerals. When he removed all the trace minerals, every type of hair became white. When he added a characteristic pattern of trace minerals, associated with a particular hair color, to a sample of demineralized hair, the color which was produced corresponded to the minerals added, and not to what the original color of the hair had been. He concluded that people inherit a tendency to concentrate certain minerals in their hair.

People who studied the effects of steroids on aging skin found that the steroids which reversed structural age changes in the skin (progesterone, testosterone, pregnenolone) sometimes restored hair growth. Occasionally, the hair that grew was pigmented. Estrogen and cortisone accelerated the structural changes of aging in the skin, but their effects on hair were not mentioned. Vitamin A has anti-estrogen effects in skin and other tissues, and part of this effect might result from its ability to promote synthesis of pregnenolone and progesterone.

Physicians have mentioned that a depigmented spot sometimes appears in the skin over an area where they have injected cortisone. The familiar association of severe stress with sudden greying of the hair also would suggest that excessive cortisone destroys melanin. The average stress caused by a particular climate would probably combine with any other factors that are involved in regions where there is more or less white hair than average.

I think oxygen wastage is a central event in aging. Just as a cut potato requires oxygen to make melanin, so do our tissues. Iron tends to keep accumulating in our tissues with aging, and iron appears to be a factor in wasting oxygen (especially in age pigment). When oxygen is deficient, iron becomes very toxic. Copper is involved in a process which restores iron to its nontoxic form.

Vitamin C in excess can contribute to the toxicity of iron, but in the right amount, vitamin C is metabolically linked with vitamin E in protecting against the toxic free radicals produced by iron. Vitamin A also functions as an antioxidant, when the amount of oxygen present is very low--which is when iron toxicity is at its worst.

Someone suspected that copper accelerated aging by causing free radical reactions, and fed a chelator (gluconic acid) to animals, and extended their lives by 10 or 12%. In his next experiment, he fed them both the chelator and extra copper, thinking he would counteract the life extension, but these animals lived 22% longer. He next fed copper alone, and found that it extended the lifespan by about 11%, so its effect was additive with that of the chelator. I suspect the chelator was removing iron, and maybe other toxic metals. At a dose about 50 times larger than normal, copper did shorten the lifespan by several percent.

Black sheep will produce white wool if they are fed an excess of molybdenum. This is supposedly caused by a displacement of copper. I think we probably accumulate too much of the wrong metals, such as iron, with aging and stress.

I had some eyebrow hairs that were pure white; when one matured and fell out, another white one would replace it. They grow quickly, and have a short life cycle, so they are nice to experiment with.

I went on a very low iron diet, eating mostly milk, with some eggs, cheese, and citrus fruit, but with very little meat for several weeks. I cooked eggs in a copper pan, to increase my copper intake and to avoid iron absorbed from an iron pan. I found a source of vitamin A without preservative, and began using large amounts of that, which I had not done for several years because of an allergy to the preservative. I increased my doses of DHEA and pregnenolone.

Usually on alternate days, I would rub vitamin A and vitamin E (sometimes with DHEA), or a solution of copper acetate, into the skin around the white hairs. Within a few weeks, the bottom of one of the white hairs had begun to darken (Figure 12.1--the hair on the right). Another hair (the center one) came out a couple of weeks later, and was darkened along about half of its length.

The third hair (on the left) came out two or three weeks later, and was all black except for 3 millimeters at the tip, which had begun growing about the time the other two were changing color. It has been about two months since I stopped cooking regularly in copper (the taste gets very tiresome), and none of the hairs has reverted to white. (The black lines were about 2 mm. apart; they allow both colors to be seen in the xerox copy.) The fine tops of the hairs end at lines 8, 9, and 7, going from left to right; each hair has a few mm. of intermediate light brown, which doesn't reproduce well.

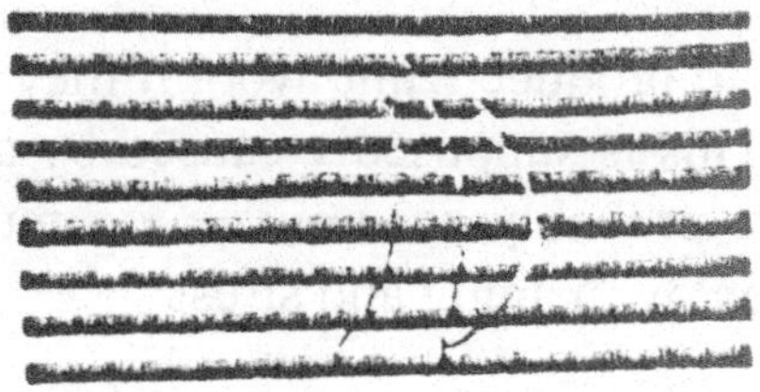

Figure 12.1 Eyebrow hairs darkening from the root up.

CHAPTER 13

ARTHRITIS AND HORMONE BALANCING

A very healthy 71 year-old man was under his house repairing the foundation, when a support slipped and let the house fall far enough to break some facial bones. During his recovery, he developed arthritis in his hands. It is fairly common for arthritis to appear shortly after an accident, a shock, or surgery, and Hans Selye's famous work with rats shows that when stress exhausts the adrenal glands (so they are unable to produce normal amounts of cortisone and related steroid hormones), arthritis and other "degenerative" diseases are likely to develop.

But when this man went to his doctor to "get something for his arthritis," he was annoyed that the doctor insisted on giving him a complete physical exam, and wouldn't give him a shot of cortisone. The examination showed low thyroid function, and the doctor prescribed a supplement of thyroid extract, explaining that arthritis is one of the many symptoms of hypothyroidism. The patient agreed to take the thyroid, but for several days he grumbled about the doctor "fixing something that wasn't wrong" with him, and ignoring his arthritis. But in less than two weeks, the arthritis had entirely disappeared. He lived to be 89, without a recurrence of arthritis. (He died iatrogenically, while in good health.)

Selye's work with the diseases of stress, and the anti-stress hormones of the adrenal cortex, helped many scientists to think more clearly about the interaction of the organism with its environment, but it has led others to focus too narrowly on hormones of the adrenal cortex (such as cortisol and cortisone), and to forget the older knowledge about natural resistance. There are probably only a few physicians now practicing who would remember to check for hypothyroidism in an arthritis patient, or in other stress-related conditions. Hypothyroidism is a common cause of adrenal insufficiency, but it also has some direct effects on joint tissues. In chronic hypothyroidism (myxedema and cretinism), knees and elbows are often bent abnormally.

By the 1930's, it was well established that the resistance of the organism depended on the energy produced by respiration under the influence of the thyroid gland, as well as on the adrenal hormones, and that the hormones of pregnancy (especially progesterone) could substitute for the adrenal hormones. In a sense, the thyroid hormone is the basic anti-stress hormone, since it is required for the production of the adrenal and pregnancy hormones.

A contemporary researcher, F. Z. Meerson, is putting together a picture of the biological processes involved in adapting to stress, including energy production, nutrition, hormones, and changes in cell structure.

While one of Selye's earliest observations related gastrointestinal bleeding to stress, Meerson's work has revealed in a detailed way how the usually beneficial hormone of adaptation, cortisone, can cause so many other harmful effects when its action is too prolonged or too intense.

Some of the harmful effects of the cortisone class of drugs (other than gastrointestinal bleeding) are: Hypertension, osteoporosis, delayed healing, atrophy of the skin, convulsions, cataracts, glaucoma, protruding eyes, psychic derangements, menstrual irregularities, and loss of immunity allowing infections (or cancer) to spread.

While normal thyroid function is required for the secretion of the adrenal hormones, the basic signal which causes cortisone to be formed is a drop in the blood glucose level. The increased energy requirement of any stress tends to cause the blood sugar to fall slightly, but hypothyroidism itself tends to depress blood sugar.

The person with low thyroid function is more likely than a normal person to require cortisone to cope with a certain amount of stress. However, if large amounts of cortisone are produced for a long time, the toxic effects of the hormone begin to appear. According to Meerson, heart attacks are provoked and aggravated by the cortisone produced during stress. (Meerson and his colleagues have demonstrated that the progress of a heart attack can be halted by a treatment including natural substances such as vitamin E and magnesium.)

While hypothyroidism makes the body require more cortisone to sustain blood sugar and energy production, it also limits the ability to produce cortisone, so in some cases stress produces symptoms resulting from a deficiency of cortisone, including various forms of arthritis and more generalized types of chronic inflammation.

Often, a small physiological dose of natural hydrocortisone can help the patient meet the stress, without causing harmful side effects. While treating the symptoms with cortisone for a short time, it is important to try to learn the basic cause of the problem, by checking for hypothyroidism, vitamin A deficiency, protein deficiency, a lack of sunlight, etc. (I suspect that light on the skin directly increases the skin's production of steroids, without depending on other organs. Different steroids probably involve different frequencies of light, but orange and red light seem to be important frequencies.) Using cortisone in this way, physiologically rather than pharmacologically, it is not likely to cause the serious problems mentioned above.

Stress-induced cortisone deficiency is thought to be a factor in a great variety of unpleasant conditions, from allergies to ulcerative colitis, and in many forms of arthritis. The stress which can cause a cortisone deficiency is even more likely to disturb formation of progesterone and thyroid hormone, so the fact that cortisone can relieve symptoms does not mean that it has corrected the problem.

According to the *Physicians' Desk Reference*, hormones similar to cortisone are useful for treating rheumatoid arthritis, post-traumatic osteoarthritis, synovitis of osteoarthritis, acute gouty arthritis, acute nonspecific tenosynovitis, psoriatic arthritis, ankylosing spondylitis, acute and subacute bursitis, and epicondylitis.

Although cortisone supplementation can help in a great variety of stress-related diseases, no cure will take place unless the basic cause is discovered. Besides the thyroid, the other class of adaptive hormones which are often out of balance in the diseases of stress, is the group of hormones produced mainly by the gonads: the "reproductive hormones." During pregnancy these hormones serve to protect the developing baby from the stresses suffered by the mother, but the same hormones function as part to the protective anti-stress system in the non-pregnant individual, though at a lower level.

Some forms of arthritis are known to improve or even to disappear during pregnancy. As mentioned above, the hormones of pregnancy can make up for a lack of adrenal cortex hormones. During a healthy pregnancy, many hormones are present in increased amounts, including the thyroid hormones. Progesterone, which is the most abundant hormone of pregnancy, has both anti-inflammatory and anesthetic actions, which would be of obvious benefit in arthritis.

There are other protective hormones which are increased during pregnancy. These include DHEA, which is being studied for its anti-aging, anti-cancer, and anti-obesity effects. (One of the reasons that is frequently given for the fact that this hormone hasn't been studied more widely is that, as a natural substance, it has not been monopolized by a drug patent, and so no drug company has been willing to invest money in studying its medical uses.) Some of these hormones also have the ability to control cell division, which would be important in forms of arthritis that involve invasive tissue growth. Since DHEA can be easily metabolized into testosterone (by the skin, for example), and into estrogen, I don't think it should ever be administered alone, without an approximately natural balance of pregnenolone and progesterone.

While these substances, so abundant in pregnancy, have the ability to substitute for cortisone, they can also be used by the adrenal glands to produce cortisol and related hormones. But probably the most surprising property of these natural steroids is that they protect against the toxic side effects of excessive adrenal hormones. And they seem to have no side effects of their own; after about fifty years of medical use, no toxic side effects have been found for progesterone or pregnenolone.

Pregnenolone is the material the body uses to form either progesterone or DHEA. Other natural hormones, including DHEA, haven't been studied for so long, but the high levels which are normally present in healthy people would suggest that replacement doses, to restore those normal levels, would not be likely to produce toxic side effects. And, considering the terrible side effects of the drugs that are now widely used, these drugs would be justifiable simply to prevent some of the toxic effects of conventional treatment. Especially during pregnancy, it would seem logical to use the natural hormones and to avoid toxic drugs.

It takes a new way of thinking to understand that these protective substances protect against an excess of the adrenal steroids, as well as making up for a deficiency. Several of these natural hormones also have a protective action against various poisons--Selye called this their "catatoxic" effect.

Besides many people whose arthritis improved with only thyroid supplementation, I have seen more than 30 people use one or more of these other natural hormones for various types of arthritis, usually with a topical application. Often the pain is relieved within a few minutes. I know of several other people who used progesterone topically for inflamed tendons, damaged cartilage, or other inflammations. Only one of these, a woman with rheumatoid arthritis in many joints, had no significant improvement. An hour after she had applied it to her hands and feet, she enthusiastically reported that her ankle had stopped hurting, but after this she said she had no noticeable improvement. (I applied it the first time, but I never saw her again, and I suspect she didn't apply it as thoroughly as I had, mixing it with oil at the time of application.)

We often hear that "there is no cure for arthritis, because the causes are not known." If the cause is an imbalance in the normal hormones of adaptation and resistance, then eliminating the cause by restoring balance will produce a true cure.

But if it is more profitable to sell powerful drugs than to sell the nutrients needed to form natural hormones (or to supplement those natural hormones) we can't expect the drug companies to spend any money investigating that sort of cure. And at present the arthritis market amounts to billions of dollars in drug sales each year. A drug company representative recently told me that a substance which cures arthritis isn't a desirable product; a good drug (from his company's point of view) is one that has to be used for the rest of the patient's life.

CHAPTER 14

THE CARPAL TUNNEL SYNDROME

When a physician chooses the most profitable diagnostic and therapeutic approaches to a health problem, it is likely to be considered fraudulent if the doctor is a chiropractor, but not if the doctor is a surgeon. If a naturopath tells you that bed rest is the best treatment for a ruptured spinal disc, surgeons will warn you about the dangers of quackery, but the research clearly shows that, for safety and efficacy, surgery is distinctly inferior to bed rest.

There are many other situations in which doing nothing, or using a more conservative treatment, is clearly superior to the standard medical or surgical treatment, but the medical industry has learned how to control public opinion by manipulating the mass media.

By consulting with thousands of women who believed they had a hormone imbalance of some sort that their physicians couldn't identify, I began to see several clusters of symptoms that responded immediately to a rational nutritional and hormonal anti-stress program. One of these clusters might be called Subtle Constellation of Absorbed-Mucoprotein-related signs and symptoms, or possibly subclinical myxedema, though neither term is likely to be widely adopted in the medical community.

Women are several times more likely than men to have "thyroid disease," simply because estrogen tends to block thyroid function. Estrogen-induced thyroid hypofunction can be compensated to some degree by various hormonal adjustments; elevated secretion of adrenalin and cortisol are common. When the compensation is inadequate, there will often be hypoglycemia and a tendency to form too much histamine. Too much adrenalin will cause cold hands and feet, too little will cause orthostatic hypotension (blacking out when you stand up too quickly) and bowel spasms, for example.

Various water-binding glycoproteins are formed under the influence of hormones or stress, but whatever proteins are in the blood, including albumin, will show up outside the blood vessels, around the tissue cells, when the blood vessels become leaky. Low thyroid, high estrogen, and high histamine are known to increase the permeability of blood vessels.

Patients who have myxedema typically have mitral valve prolapse, and at autopsy it can be seen that the valve is thickened into a jelly-like mass. Many women with the premenstrual syndrome have a mitral valve heart murmur premenstrually, but not at other times of the month. The jelly can be formed and removed fairly quickly.

Old textbooks on the thyroid gland often listed emphysema as a symptom of myxedema. When rats are given a large injection of estrogen, the oxygenation of blood in their lungs is sharply decreased in less than an hour. Although it is not common to test oxygen diffusion in humans, I know of two women in the same family who showed very poor pulmonary oxygen diffusion after they were put on high doses of estrogen. (Since estrogen inhibits thyroid secretion, and hypothyroidism is associated with elevated estrogen, they should be considered together.) Leakage of proteins from capillaries in the lungs is probably responsible for the decreased diffusion, by thickening the layer through which the oxygen must diffuse.

When myxedema exists in childhood, the cartilage in joints swells, and is deformed, causing a characteristic knock-kneed appearance. In a milder form, the swelling can cause joint pain that doesn't involve the characteristic inflammation of arthritis. In a mild form, the calf muscles tend to swell after prolonged activity;

the "growing pains" that are so common around the beginning of puberty are probably the result of a temporary hypothyroid edema of the leg muscles.

When a tendon swells, it is sometimes the result of a local injury or of a particular overuse of a muscle, and in these cases local treatment can sometimes produce a permanent cure. But when the problem keeps recurring, or keeps showing up in different areas, there is probably a general hormonal problem.

When a tendon in the wrist swells, it can cause numbness in the hand, by pressing on a nerve which passes through the carpal tunnel with some tendons. The tunnel is formed by a ligament that holds the tendons in place, and swelling of the ligament itself can contribute to compression of the nerve. Even the connective tissue that forms the nerve sheath itself can swell. Many people with undiagnosed hypothyroidism complain that they "have poor circulation," and that their hands and feet go to sleep easily. These are two separate (but related) problems that can be relieved through a nutritional and hormonal anti-stress program.

Low thyroid people often have cold hands and feet, and they often have nerves that are oversensitive to compression. Poor oxygenation is involved in edema, both as cause and consequence. (I discuss this in more detail in "A Unifying Principle," chapter 7.)

Thus, swelling can cause compression and nerve injury in a way that is exacerbated by certain postures or repetitive nerves and arthritis commonly follow treatment of the carpal tunnel syndrome by surgery or local cortisone injections. Glaucoma, a swelling of the jelly inside the eye, seems to result from a hormonal imbalance, since it can be relieved quickly by a natural hormone. Graves' disease, in which the muscles behind the eyes swell, can also be aided by similar treatments. In addition, "epilepsy" and symptoms ascribed to "multiple sclerosis" have been relieved by natural anti-stress and anti-edema hormones.

CHAPTER 15

THE PREMENSTRUAL SYNDROME

For at least 80 years, a syndrome of premenstrual tension has been recognized and considered to be the result of an ovarian hormone imbalance. As early as 1912, Armour & Co. was selling desiccated corpus luteum for use in cases of ovarian failure, and said that it "prevents nervous symptoms accompanying" menstrual abnormalities. (*New and Nonofficial Remedies*, 1912, Supplement, pages 158, 160.) It was also used to treat obesity and other physical conditions sometimes associated with "ovarian deficiency."

By the 1930s, when the specific ovarian steroids were identified, there was a clear consensus that estrogen and progesterone were involved in the syndrome, but there was some confusion over the question of whether the cause was high estrogen or low progesterone. It seems to have been French endocrinologists who pointed out that either high estrogen or low progesterone can cause the typical cyclic premenstrual symptoms, since it is the ratio of estrogen to progesterone which determines the nature of the physiological response. By the mid-1930s pure progesterone was on the market in the United States and Europe, and was widely used to treat problems of pregnancy, menopause, dysmenorrhea, and the premenstrual syndrome.

Although it has been only in the last 20 years that animal studies of the behavioral effects of the steroid hormones have become subtle enough to explain their role in emotions, depression, aggression, etc., the more physical dangers of excessive or unopposed estrogen were clearly recognized by the late 1930s and the early 1940s: It was, for example, known to cause cancer, infertility and miscarriages, excessive blood clotting, edema and a shock-like stress reaction. Other toxic effects of estrogen--seizures, birth defects, toxemia of pregnancy, increased intraocular pressure, atrophy or necrosis of various tissues, etc.--continued to be discovered, while the opposing steroid, progesterone, was found to offer protection against these, and still other, toxic effects.

Although the premenstrual tension syndrome, in its behavioral and subjective aspects is now recognized as a significant (and treatable) problem affecting a very large proportion of women in their reproductive years, another very important issue is that the hormone imbalance which causes the premenstrual discomfort and disability also predisposes the sufferer to serious organic problems.

While both animal studies and human studies have shown that natural and synthetic progestins protect against estrogen-induced cancers, and some other diseases, only natural progesterone is recognized as having general value in the treatment of the premenstrual syndrome. This is apparently because natural progesterone has a broad spectrum of intrinsic effects, and is easily metabolized into other active hormones as needed, that is, it is a basic precursor substance. It has been tested in isolation on practically every type of tissue, where it seems to have a stabilizing action: nerve, smooth muscle, cardiac muscle, epithelial cells, cells of the immune system, etc. It is antispasmodic, anti-inflammatory, anti-mitotic in certain tissues, and (at high concentration) anesthetic. Its use before and during pregnancy has been associated with a lower than normal incidence of birth defects.

Although natural progesterone has a broad spectrum of action and is remarkably free of side effects, the use of natural progesterone has been limited by the expense and inconvenience of the forms commonly used. Some women continue to use suppositories, because they do offer some relief of symptoms, even though they are very messy, and they are necessarily expensive, since only 5% or less of their progesterone content can be absorbed.

Injectable progesterone is both inconvenient and expensive. The form commonly used contains 10% benzyl alcohol, as a solvent and as a local anesthetic, since progesterone is not sufficiently soluble in ordinary vegetable oil, and can be painful when injected as an aqueous suspension. However, benzyl alcohol has significant toxicity to nerves, and can produce anaphylaxis.[1] Benzyl alcohol, in contact with tissue, combines with water, causing progesterone to be deposited as crystals. Even if oils needed a "bacteriostatic agent," the quantity used with progesterone has no justification except as a solvent, but it is not described as such. The fact that injected progesterone has been beneficial, in spite of the neurotoxicity and instability of the solvent, is another illustration of the protective action of progesterone.

Micropulverized progesterone is expensive when taken orally, since it is poorly absorbed, and when absorbed it is quickly inactivated by glucuronidation, since it is exposed to enzyme action in the wall of the intestine itself and in the liver, if the individually absorbed molecules get that far before being inactivated.

Substances, even including peptides, which are absorbed by the chylomicron pathway, reach the general circulation without being exposed to the inactivating glucuronide-transfer enzymes of the intestine or liver. People often speak of "avoiding the liver on the first pass," but in fact chylomicrons pass through the liver many times before they are destroyed; after an hour, 10% of the chylomicrons are still circulating. If progesterone is taken orally in oil in a truly solvated state, in a monomolecular dispersion, it will enter the blood via the protected chylomicron route. Merely mixing finely powdered progesterone crystals with oil will not allow it to enter as chylomicrons; the size of chylomicrons is very small. They are usually about 4000 Angstrom units in diameter.

Fats taken orally are almost 100% absorbed by the small intestine. Like other hydrophobic substances, progesterone will be absorbed by adipose tissue and by the brain, but, unlike other steroid hormones, it also tends to be absorbed by red blood cells. Typically, the concentration of progesterone in red blood cells is twice that of the serum.[2] (And the brain contains a still higher concentration.) These intracellular reservoirs of progesterone tend to prolong the elevated blood levels resulting from the absorbed chylomicron-progesterone, so that the observed hormone levels after a single oral dose are much more stable than are the triglyceride levels after a single fatty meal. Typically, after a single oral dose of 100 mg. of perfectly dispersed progesterone, a post-menopausal woman's serum progesterone level will still be in the "normal luteal phase range" 24 hours later. The difference in efficiency--and therefore in cost--between this form and the various crystalline dispersions is very large.

Natural progesterone therapy in the form described above remains the most direct and general treatment for PMS; thyroid and vitamin A, and other normalizers of mitochondrial respiration, such as red light and short chain saturated fats, act in the same direction. Fiber, bowel stimulants and antispasmodics, and broad spectrum bowel "disinfectants" such as flowers of sulfur, sodium thiosulfate, and short and medium-chain saturated fatty acids, can contribute to alleviation of premenstrual symptoms (as discussed in other articles). Penicillin can also tentatively be added to the rational therapies for PMS, as a flora-normalizing one-time or occasional treatment, though I think the other methods are so effective that any need for penicillin would be rare.

My first suggestion for someone with PMS is to avoid thyroid suppression (diet; darkness and endurance exercise should be avoided), and to use my carrot salad recipe: Grated carrots, vinegar, coconut oil and salt are the essentials, garlic and olive oil are optional. Acetic acid and fatty acids released from the coconut oil act at different levels, and the carrot fiber is a timed-release system which also binds toxins and stimulates the bowel; the salt spares magnesium and tends to inhibit excessive prolactin release. Use of a balanced thyroid (Armour or Proloid) supplement can help to restore normal thyroid function by breaking the cycle of stress. If these methods aren't enough, progesterone or pregnenolone can be supplemented. Both of these help to normalize thyroid function, and both have the "catatoxic" function of protecting against a wide range of toxins.

REFERENCES

1. Feasby, T. E., et al. (1983). Neurotoxicity of bacteriostatic water. *N Engl. J. Med, 308*(6), 966-7.

Kimura, E.T., et al. (1971). Parenteral toxicity studies with benzyl alcohol. *Toxicol. Appl. Pharmacol., 18*(1), 60-68.

Grant, J.A., et al. (1962). Unsuspected benzyl alcohol hypersensitivity. *N Engl. J. Med., 306*(2),108.

2. Mulder, E., et al. (1972). Metabolism of free and conjugated steroids by intact and haemolysed mammalian erythrocytes. *Biochem. Biophys. Acta., 260*(2), 290-297.

CHAPTER 16

RESTORING FERTILITY

Since my study of the role of oxygen in fertility, between 1970 and 1972, I have concentrated on working out the implications of what I observed, and have written very little about the treatment of infertility. But every time someone would mention that they were unable to get pregnant, I would explain what could be done, and they would get pregnant, usually within a few weeks. The trick is simply to normalize the use of oxygen, by normalizing everything related to hormones and nutrition.

AGE LIMIT CONSIDERATIONS

Several years ago, a friend of mine had a healthy baby a few months after she had begun using small supplements of hormones, and several years after her early menopause. I wondered whether age or the absence of menstruation might be almost irrelevant to fertility, if the right balance and sequence of hormones could be achieved.

A gynecologist I knew had been giving his wife progesterone since she was about 40, and at 60 she was youthful looking, and still menstruating. I have known of many women who still menstruated at 55, and--since diet and light and stimulation powerfully affect metabolism of hormones--it is easy to imagine that natural conditions could sometimes help to preserve fertility beyond that age.

A grave monument in St. Peter's Church in East Oxford records the death of a woman in childbirth at the age of 62.

The old medical and obstetrical journals have many reports of women who had children long after menopause.

Depasse (*Gazette Medical de Liege*, Oct. 1, 1891) described a woman who had a child at the age of 59 years 5 months, and who suckled the baby, and weaned it on her 60th birthday. Her menopause had occurred 10 years before the pregnancy.

The famous American physician, Benjamin Rush (in *Medical Inquiries and Observations*, Philadelphia, 1805) described a woman of 60 who had a child.

Wallace (*Philosophical Transactions of the Royal Society of London*, vol. xxii, Page 543) reported a woman on the Isle of Orkney who had a child when she was over 60.

There have been a few reports of children borne by women of 70 (*L'Union Medical du Canada, Montreal*, Dec. 3, 1881), and older (Mayham, in *Medical Standard*, Chicago, Jan. 1891).

There are a few accounts of even older women who had healthy babies, and occasionally a calcified fetus is found in the uterus of an extremely old woman during autopsy. The assumption that the "child-bearing years" end around the age of 50 has led to the conclusion that such fetuses have remained "in utero" for decades, but no one knows how long the calcified fetus can persist in the uterus.

Taboos undoubtedly influence the willingness of pregnant old women to tell their true age. Similarly, our cultures have tended to blame women for all reproductive problems. (In *Nutrition for Women*, I discussed some of the evidence that birth defects are often the father's contribution, but since then the evidence in support of that view has become much clearer. Male geneticists used to have a little spiel which "explained" why women were responsible for Down's syndrome, etc., but it actually explained nothing.) "Everyone knows" that old men can produce children far beyond the age at which women become infertile. Our culture honors old men who can reproduce, and so the claims of fatherhood at a great age are likely to be more numerous than the claims of mothers that they are very old. I knew a man in his 80s whose pretty wife of about 30 had a baby, but not everyone believed he was the father. However, when a woman has menopause at the age of 47, marries a 22 year old man when she is 52, and has children when she is 53 and 54 (Cachot, in *Pacific Medical and Surgical Journal*, vol. xxvi, page 394, 1883-4), there can't be the same sort of doubt. In the latter case, one could assume that a young husband optimized the couple's fertility.

Since both parents are involved in reproduction, couples that want to have children must consider the health of both. Thyroid and pregnenolone and vitamin E are as important for male fertility as thyroid and progesterone and vitamin E are for female fertility. (For example, supplementary thyroid and pregnenolone can raise a man's sperm count, by overcoming the effects of stress.)

THE MECHANISMS OF FERTILITY

Since it helps to have information from blood tests, physicians are often involved, and it is important that both the woman and her doctor be aware of what the other is trying to do. For example, the thyroid hormone, triiodothyronine (T3), is a key element in producing the energy needed for pregnancy to begin, but if there isn't enough cholesterol circulating in the blood, it will be impossible to synthesize enough progesterone (and other steroids). Too much carotene, too little vitamin A, not enough magnesium or sodium, and too much cortisol are commonly overlooked factors in infertility.

It isn't uncommon for women to experience monthly periods several months into pregnancy; even when there isn't continued monthly bleeding, the cervical mucus test will often show that there is a cyclic monthly shift toward estrogen's dominance. These are times when the probability of a miscarriage is increased.

The most common time for miscarriages is at the ninth week. At this time (just as the placenta is about to take over the bulk of hormone production from the ovaries) there is normally a brief decrease in the ratio of progesterone to estrogen. The decreased ratio of progesterone to estrogen is characteristic of normal menstruation, and when it occurs during pregnancy, there will be either some degree of bleeding or a change in the structure of the mucus, if not a miscarriage. Estrogen lowers the availability of oxygen, and progesterone increases it. Bleeding is caused by changes in the blood vessel (spasm followed by dilatation) promoted by nerves, under the influence of estrogen and the resulting oxygen-deficiency. The rate and quality of fetal development is influenced by these hormone balances.

In animal experiments, an appropriate dose of estrogen terminates pregnancy at any stage. It is clear that lack of oxygen caused by estrogen dominance is a major factor in miscarriage at any stage. The six-day-old embryo suddenly becomes dependent on oxygen, just at the moment of readiness for implantation. Only a small excess of estrogen is needed to prevent implantation of the embryo, because progesterone is produced in relatively small amounts at that early stage, but even late in pregnancy a dose of estrogen sufficient to override the effects of progesterone can destroy the fetus.

In smaller doses, estrogen damages the fetus without killing it, by stopping brain growth, for example. In the later stages of pregnancy, any slight oxygen deficiency causes an excessive use of glucose, as a partial compensation, and the resulting hypoglycemia can cause symptoms in the mother while damaging the fetus. Morning sickness, temporary blindness, fainting, and seizures are common effects of low blood sugar, especially after the fifth month, when the baby's brain is growing rapidly and consuming increased amounts of glucose.

Adequate protein, glucose, and sodium to maintain blood volume will prevent most of these problems of later pregnancy, unless the hormonal imbalance is very bad. Milk and fruit juice taken several times a day are usually helpful. Body temperature and pulse rate should be monitored, to see whether the metabolic rate is approximately normal. Sometimes, a single dose of progesterone can normalize a pregnancy, but both thyroid and progesterone should be considered for improving fertility and preventing miscarriage. Women who used progesterone before getting pregnant had babies with a very low incidence of congenital defects (roughly 1/10 the "normal" rate), and women who used sizable amounts of progesterone during pregnancy had outstandingly superior babies. The things which sustain fertility, in its simple sense, also are the things which optimize the life of the child.

PART III

REGENERATING KNOWLEDGE

"As the true method of knowledge is experiment, the true faculty of knowing must be the faculty which experiences."

W. Blake

Our organism has requirements that must be recognized, and one of these is the need to know, to have intellectual adequacy. Knowing takes energy, and our experience is influenced by our biological state. We can improve our developmental conditions, socially and materially, to bring our species up to a higher level of mental energy. In this section, I discuss some of the biological, social, and ideological aspects of the problem--the problems of how to overcome inertia and foster creative knowing.

CHAPTER 17

YOUTH, ENERGY, AND REGENERATION

Being alive is good for you. But our culture is saturated with arguments to the contrary--that it is life which kills us, and self-denial which sustains us. It is always easier to blame the victim than to search for the real cause of a problem. In this chapter I will give some of the arguments in favor of life and more life, or rather, some of the evidence from which those arguments can be made.

Fifty years ago, Clyve McCay reported that restriction of calories in the diet increased animals' longevity. At that time, many biologists believed in an innate and universal "metabolic potential," according to which each gram of living tissue could produce only a certain number of calories of energy metabolically, beyond which it must decline and die. In Raymond Pearl's version, this limit was governed by heredity, and was originally demonstrated by showing that different individual organisms could survive total energy deprivation for different lengths of time; slow metabolism postponed starving. (Obviously, starvation has no logical connection with aging.) Later, Pearl argued that more intense activity accelerated aging.

There are now many people who argue that a low metabolic rate, a low body temperature and slow heart beat indicate that you should live a long time: "your heart can beat only so many times." Most of these people also advocate "conditioning exercise," and they point out that trained runners tend to have a slow heart rate. (Incidentally, running elevates adrenalin, which causes increased clumping of platelets and accelerated blood clotting. Hypothyroidism--whether pre-existing or induced by running--slows the heart, raises the production of adrenalin, and is strongly associated with heart disease, as well as with high cholesterol.)

Early in the century, Alexis Carrell showed the importance of environmental factors in the aging process, by keeping chicken cells alive and functioning for many years beyond the chicken's normal lifespan. In 1927, German researchers reported that a fat-free diet prevented the occurrence of spontaneous cancers in rats. Since, a little later, other workers found that the elimination of unsaturated fats from the diet not only prevented cancer, but also caused a large increase in the metabolic rate, it might have been possible to conclude that it is not living which kills us, but something in the environment. Some people did draw that conclusion, but research funds go mainly to product-oriented research, and "the environment" has been hard to package as a product.

Carrell's work involved frequent changes of the fluid that the chicken cells were growing in. Later, Hans Selye did an experiment in living rats that seems to relate to the importance of frequent changes in the growth medium. Selye implanted glass tubes under rats' skin, and found that they were soon encapsulated by fibrous tissue, and that a filament of tissue then grew down the tube, joining the membranes at the ends. When left undisturbed, this filament, isolated from the circulatory system, underwent very rapid aging.

But when Selye drained the fluid at frequent intervals, the age-changes were prevented, and even when the rats died of old age, the filaments still appeared to be young. Another gerontologist, who had an extremely old and decrepit dog, removed, a little at a time, the dog's blood, discarded the serum, and replaced the cells suspended in a saline solution; the dog appeared to be rejuvenated, and regained all its normal functions. The blood serum from young animals or people supports better cell growth in vitro than does the serum from older individuals.

Two clear differences have been found between old blood and young blood. The albumin in old blood is in a more oxidized state. (I think it was the famous gerontologist, Verzar, who first reported this.) Although, at least in aging humans, there is much less oxygen in the blood, something causes the albumin to be in a more oxidized state in older blood. The other distinct feature of older blood might also seem paradoxical at first: the red blood cells are younger. That is, in an old individual, the red blood cells are more fragile--possibly from being more quickly damaged from oxidation--and are replaced sooner, and so, on average, they are many weeks younger than the cells in a healthy young individual.

Neither of these features is paradoxical. Poor oxygenation is a stress, and causes the waste of glucose and compensatory mobilization of fat from storage, and the relatively reducing environment in the cytoplasm causes the mobilization of iron from storage, in the toxic reduced (ferrous) form. Products of the peroxidative interaction of iron with unsaturated fats are evident in the blood (and other tissues) during stress, and especially so in older animals.

Several things relating to calorie-restricted diets are not generally known. First, in 1987, Schroeder found that the removal of toxic heavy metals from the diet had the same effect as caloric restriction. Second, although underfed animals grow more slowly, their metabolism is not necessarily depressed. (In fact, animals on a low protein diet have a higher rate of oxygen consumption than do the animals that eat a more normal diet.)

Many older studies found that rats which were deficient in the "essential fatty acids" had the signs of hyperthyroidism, and were in a hypermetabolic state. The avoidance of oxidatively toxic heavy metals, and the maintenance of respiration, with an absence of the highly peroxidizable unsaturated fats in the diet (and a lower level of them in the storage tissues) would probably make the animals tolerate stress better (EFA deficient mitochondria are more resistant to oxidative injury, and vitamin E prevents many stress-associated problems), and might inhibit the age-related oxidative changes in serum albumin, red blood cells, and other tissues.

A third finding related to dietary restriction concerns protein turnover. One of the basic metabolic changes in aging is slowing of the rate of protein turnover in cells, and it appears that dietary restriction enhances the protein turnover rate in aging animals. I think it is likely that unsaturated fats and the amino acid, cysteine, both contribute to the age-related retardation of protein metabolism, but I don't think this has been directly tested yet.

(In reading the published research on dietary restriction and the effects of unsaturated or saturated fats, it is important to pay careful attention to the actual composition of the diet. The cancer rate, for example, usually reaches a maximum and levels off at a corn oil content of less than 10% of calories, but this content is often called a "low fat" diet. If a normal and low calorie diet both contain 6% corn oil, which is clearly carcinogenic, the animals which eat less food will eat less corn oil, and will also tend to burn it more completely for energy, so there will be a major difference in the toxic effects of oil. Lard is often the same as soy oil in linoleic acid content, yet it is too often described as a "saturated fat." I read a publication on the effects of a "low fat" diet, which contained 20% corn oil; sometimes I feel there is a deliberate attempt to mislead when language is used so arbitrarily.)

Besides the observation of greater oxygen consumption in the low fat animals, and high protein turnover in calorie restricted animals, there are observations in a variety of organisms associating a higher metabolic rate with greater longevity. While most longevity studies of flies involved altering the temperature of their environment, studies of differences of metabolic rate at a given temperature have in several cases found greater longevity in the high metabolizers. A study of 18 strains of mice found a clear association between a higher metabolic rate and greater longevity.[1] Recent studies (e.g., Joseph Graves') are showing similar associations in insects.

Sacher popularized the idea that a larger brain is associated with a longer life span, and others more recently have refined the idea in connection with body size, index of cephalization, and metabolic rate. For example, M. A. Hoffman (Energy metabolism, brain size, and longevity in mammals, *Quar. Rev. Biol. 58*(4), 495-512) said "...it has been shown that the ratio of cortex-to-brain metabolic rates is independent of body size, and increases with the evolutionary level of brain development." (Figure 17.1) It is interesting to consider that birds generally live longer than mammals of the same weight, though their brains are usually smaller, and their body temperature is several degrees warmer. Among birds, parrots and other relatively large-brained birds have an extremely long life span, compared with mammals of the same body size. For example, a rat lives about 2 years, at 98 degrees F., and a medium sized parrot lives about 70 years, at 104 degrees. According to H. Rahn ("Time, energy, and body size," chapter 16 in *Environmental Physiology*, editors C. V. Pagenelli and L. E. Farki, Springer-Verlag, 1988), the total energy per gram of tissue per life span is about four times higher in passerine birds than in mammals.

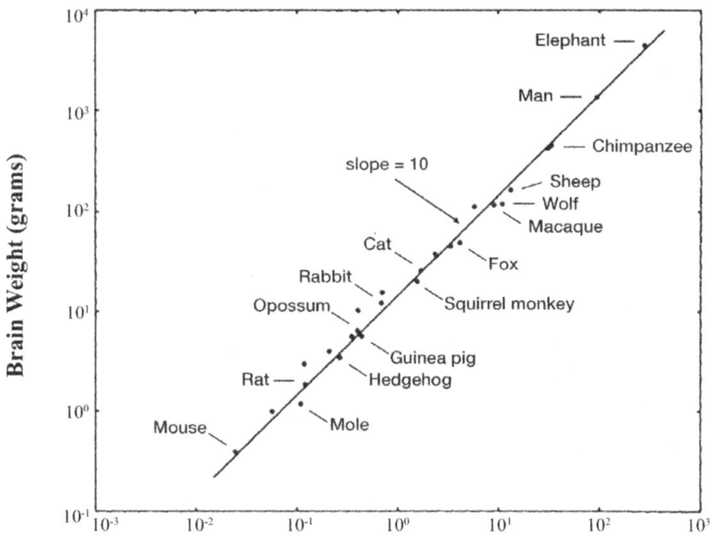

Figure 17.1 Hofman, M. A. (1983). Enlargements of Brain Size Keep Pace with the Availability of Energy. *The Quarterly Review of Biology*, *58*(4).

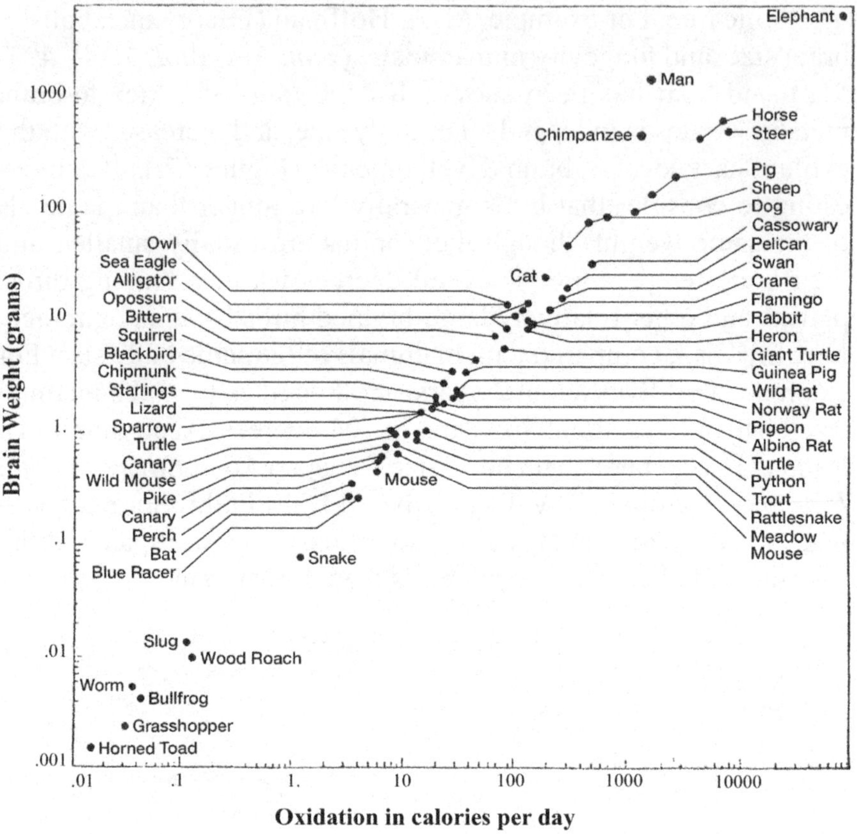

Figure 17.2 G. Crile: In relation to metabolism, the weight of the brain shows a law.

Undoubtedly, he meant us to infer that these "uncivilized" people fell on the same line as animals, because their brains lacked the cultural quiet parts. My interpretation is that they ate traditional diets, rather than the grain-based diet of the United States and Europe which suppresses brain metabolism, in the same way that the "normal" animal diets suppress metabolism (and decrease longevity). Many researchers (before the late 1940's) found that about 40% of the people in the U.S. showed evidence of deficient thyroid function (low oxygen consumption and high serum cholesterol), and benefitted from taking a thyroid supplement. Cancer, heart disease, and high susceptibility to infections were found to be extremely prevalent in populations with subnormal thyroid function.

Low thyroid function, relative overfeeding, and the presence of unsaturated oils in the diet are known to accelerate sexual maturity. Early sexual maturity has been associated with premature aging and early death. Fish, octopuses, mice, humans, and plants offer examples in which reproductive maturity initiates the aging process. Although it used to be said that "hot tropical" people had early puberty, and "cold northern" types had late puberty, the best available data contradict that opinion. The oldest averages for the occurrence of puberty occur in tropical regions. (Figure 17.3)

Figure 17.3 Age at Puberty: Caribbean: 15.1 years; Central Africa: 16.5 - 17 years; New Guinea: 15.5 - 18.4 years.

Mere calorie restriction can delay puberty (and this usually means a low fat diet, for poor people in the developed countries) as can be seen in data from Appalachia; late puberty, accompanied by very low birth weight for babies, is the typical pattern of poverty. Given enough fat (especially vegetable oil, including that in beans and corn), harsh conditions can probably cause earlier puberty, but I don't know of any clear evidence on this subject.

In the tropical regions where puberty doesn't occur until about the age of 17, it seems that some of the heaviest birth weights occur, and high birth weight generally indicates good maternal nutrition. (Figure 17.4) These tropical areas generally coincide with areas where coconuts are produced abundantly, and are used for food. Animals fed coconut oil are lean, have delayed sexual maturation, and have offspring with larger brains than animals fed similar amounts of unsaturated oils. Since coconuts and related fruits contain sterols, their use might have a direct hormonal action, in addition to the effects of the easily metabolized high energy short and medium-chain fats, which support an efficient and intense respiratory metabolism. The dietary hormones are probably responsible for the high metabolic rate of the Eskimos, too, except in that case glands, skin, and brains are regularly eaten. Tropical people generally consume all parts of their food animals, too (except for vegetarian cultures such as Hindus, which use only milk and butter from their cows), though the Eskimos are unique in having, traditionally, a pure meat diet.

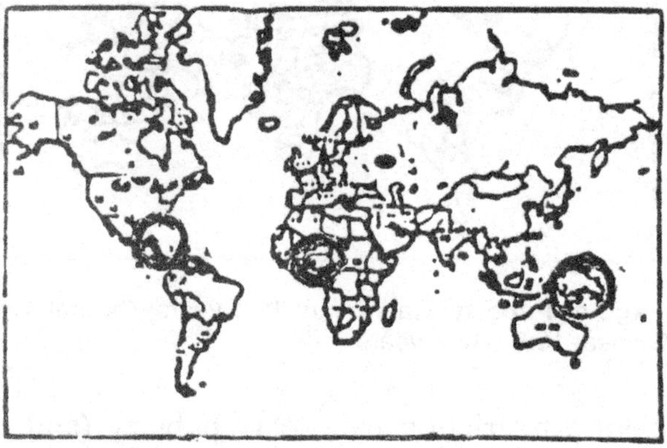

Figure 17.4 Birth Weight: Caribbean, Central Africa and New Guinea have the heaviest birth weights in the world: 3.4 kgs.

W. Donner Denckla suggested that there is a "death hormone" in the pituitary gland, which appears at puberty and initiates the process of aging by suppressing the use of oxygen. He maintained that merely providing thyroid supplement wouldn't protect against it, and that it was a distinct hormone, although it tended to appear in tissue extracts in association with prolactin and growth hormone. Although I think there is still a lot to be learned about the pituitary hormones, I don't think Denckla discovered anything except puberty.

The sex hormones--especially estrogen--are closely involved with respiration. My dissertation, "Age-related oxidative changes in the hamster uterus," surveyed some to the same information. Estrogen and prolactin do many things to interfere with respiration. L. C. Strong, who developed strains of mice with high estrogen and a tendency to die of mammary cancer, found that early sexual maturity was associated with a shorter lifespan. Similar observations have been made in humans.

Rather than supposing that we contain a system for self-destruction, I suspect that reproductive maturity appears at a time when the organism is experiencing a generalized, life-threatening metabolic stress--namely, aging--and that the metabolic slowing which begins at puberty signals the appropriateness of reproduction, and makes the organism more dependent on the glucocorticoids (regulated by the pituitary). The accumulation of iron, unsaturated fats, and other toxins, I think, are major signals for puberty. (Specifically, factors which inhibit cytochrome oxidase.)

Marion Diamond, who studied the effects of stimulation on rats' brain development, found that pregnancy or progesterone treatment--like freedom and stimulation--caused the brain to grow, and estrogen--like stress--caused it to shrink. The larger-brained animals had smaller faces, and the confined, stressed rats had smaller brains and bigger faces. (I see no reason to doubt that human proportions are similarly influenced by prenatal stress; Franz Boas, for example, found that immigrants' childrens' head shapes were similar to head shapes in the new country, and unlike their parents.') These changes are similar to the differences between small-faced babies and large-faced adults. The observation that the babies of one species resemble the adults of a more highly evolved species has led to the idea of evolution through neoteny (or paedomorphism or juvenilization), in which juvenile traits are preserved for a longer time in the offspring, and eventually persist even after reproductive maturity is reached.* The same terms are sometimes used in describing plants, as well as animals.

Some feminists have objected to the characterization of female traits as "paedomorphic," considering it as another attempt to deny women equal social status with men. It was common for Ashley Montague's book, *Natural Superiority of Women*, to be condemned by "feminists" as "patronizing." Nevertheless, I think it is a mistake to argue for a kind of biological relativism (analogous to cultural relativism) which avoids the idea of a possible evolutionary significance in certain biological features.

If a certain substance, such as progesterone, closely reflects the quality of environmental support and is intimately involved in realizing and sustaining our most characteristically human evolutionary features, I don't think we should deny its importance just because of its greater importance in women than in men. Progesterone's effects are "neotenous," in the sense of prolonging youthful traits. Women have several "neotenous" features relative to men, including a bigger ratio of brain to lean body mass, a smaller face-to-cranium ratio, differences in voice and body hair, less aggressiveness, and greater adaptability. (In spite of the people who teach assertiveness, I think high adaptability and low aggressiveness are characteristic human and primate traits, which are typical of infants, and are likely to represent our species' future.)

H. G. Wells wrote that:

> *"The creature called Homo sapiens emerges from among the earlier Hominidae, very evidently, as another of these relapses of the life-cycle towards an infantile and biologically more flexible form which have played so important a role in the chequered history of living things.*

He is not the equivalent of the clumsy adult Heidelberg or Neanderthal man. He is, in his opening phase, the experimental, playful, teachable, precocious child, still amenable to social subordination when already sexually adult. The ever-changing condition of life had less and less tolerance for a final gross, overbearing adult phase, and it went out of the cycle. . . . . .the now unnecessary primordial adult Homo, for all effective purposes, faded out, leaving as his successor the childlike Homo sapiens, who is, at his best, curious, teachable and experimental from the cradle to the grave.

...It is possible that the mass of contemporary mankind may not be as readily accessible to fresh ideas as the younger, more childish minds of earlier generations; and it is also possible that hard imaginative thinking has not increased so as to keep pace with the expansion and complication of human societies and organizations. That is the darkest shadow upon the hopes of mankind.

There is no way out for Man but steeply up or steeply down."

A high level of metabolic intensity is characteristic of both young plants and young animals under conditions that are ideal for their life, meeting their energy and substance needs abundantly. Both plants and animals are able to adapt--phylogenetically and ontogenetically, i.e., through both transgenerational and developmental changes--to marginal conditions of energy and substance availability.

The term "paedomorphic" has been applied to plants which preserve the "youthful" pre-reproductive forms for extended periods. Premature reproduction can be produced by stress, or to look at it from another angle, the conditions which prolong the growth state can be thought of as stress-free conditions. (R. Criddle and L. Hansen, in *Plant, Cell, and Environment*, report that a plant's metabolic rate, measured as heat production, can predict its future growth rate. Jerry Barnes, a tree developer in Cottage Grove, Oregon, discovered several years ago that the intensity of the metabolic activity even in seeds can be used to accurately predict the future performance of the tree, cutting many years out of the process of tree improvement. The story of how his work was received at the state university is interesting; he did some quick lab tests on seed provided by the university, and they confirmed that he had correctly chosen the seeds which produced the best trees in their 13 year study, but they simply rejected his result as not conforming to accepted ideas.)

According to Jerry Barnes' theory, trees become able to move into less favorable niches by accumulating "restraint genes," which limit their growth, but prevent death by making them able to tolerate marginal conditions.** The theory of Criddle and Hansen seems to be similar, in looking for a stable metabolic rate when the plant is subjected to stress.

Upward adaptiveness, which is typical of large-brained animals and of plants with a high metabolic rate, allows the organisms to find more expansive niches by living at a higher energy level. This process obviously places great importance on an environment which can provide abundant energy and the necessary nutritional substances.

The fact that steroids seem to be made by all organisms except viruses, and are often identical in very different organisms (e.g., estrogen in yeast is exactly the same as human estrogen), has led me to experiment with "animal" hormones, such as progesterone, on plants. I think the systematic effects on "gene expression" and biochemical functioning that steroids are known to have in animals can also be seen in plants.

Producing energy abundantly, and using it efficiently: This seems to be an important effect of certain steroids. It might amount to a kind of by-passing of the "restraint genes." The kinds of changes which can be induced by hormones can also be inherited.[2]

If we optimize the known factors which improve energy production (red light, short-chain and medium-chain saturated fats, and pregnenolone, for example), to the extent that our metabolism resembles that of a ten year old child, I don't think there is any reason to suppose that we wouldn't have the regenerative, healing abilities which are common at that age. I suspect that both brain growth and remodeling might proceed indefinitely.

There is evidence that the cranial volume grows into old age, and also that the DNA content of the brain keeps increasing during adulthood. Presumably, this is only the result of an increase in connective tissue cells, but the certainty many people have that we can't grow new brain cells at any age is unfounded. Heart cells can now be made to multiply *in vitro*, and probably *in vivo*.[3] There is also some evidence for regeneration of brain cells *in vivo*.[4] and in birds brain cells have been found to grow new each spring in a part of the brain concerned with their singing, under the influence of a hormonal change. If the brain normally added a significant percentage of brain cells during adult life, the chance of finding a dividing cell by microscopic examination would be so small that we can't take the present absence of evidence for an increase of brain cell number during adulthood as having any meaning. The old doctrine, which claimed that "100,000 brain cells die every day," is now recognized to have no basis in fact.

For example, Bigl, *et al.* (*J. of Gerontology 33*, 172-180, 1987, citing the work of H. Haug, "Are neurons of the human cerebral cortex really lost during aging? A morphometric examination," in Traber and Gispen, *Senile Dementia of the Alzheimer Type*, Springer, 1985) say ". . it was demonstrated that an age-dependent embedding-shrinkage of brain tissue together with a secular acceleration in body size may significantly affect morphometric evaluation of neuronal densities during aging, and the view of a general neuronal loss during aging had to be corrected."

Rather than a programmed or random continuous loss of cells, when atrophy of the brain occurs, it seems to be caused by specific conditions, such as stress with prolonged exposure to glucocorticoid hormones. While estrogen is still sold for "estrogen replacement therapy" of menopausal symptoms, the typical symptoms of menopause (such as osteoporosis and hot flushes) can be produced by cortisol, and it is clear that these symptoms coincide with loss of the protective steroids such as progesterone, and not with a deficiency of estrogen.

The skeletal changes (shrinkage, curving of the back, moving forward of the lower jaw) which are so characteristic of old age in humans, also occur in other animals in aging and under the influence of the stress hormones. Since the protective hormones depend on the ability of mitochondria to convert cholesterol into pregnenolone, it is clear that damage to mitochondria will affect our supply of protective hormones at the same time that our energy supply is failing, forcing us to shift to the atrophy-producing stress hormones, including cortisol. Simple factors which protect the mitochondria are known to have profound therapeutic effects. At a certain point, I think we will understand mitochondrial protection well enough to prevent and cure the basic pathologies of aging. The Mayans and Eskimos studied by Crile produced 25% more biological energy at rest than people in the U.S. and Europe. They are culturally and nutritionally very different from each other, but they have enough in common to make them very different metabolically from the EuroAmerican culture. What they have in common is possibly something as simple as the absence of thyroid-inhibiting substances in their diet.

I think it is likely that our present knowledge of mitochondrial protection could give the average adult about a 50% increase in biological energy. To go beyond that level, it might be necessary to start at an earlier age, to allow body proportions to develop appropriately. For example, to handle a doubling or tripling of basal metabolic activity, a lung capacity of four or five liters would probably be needed. Like people who grow up in the high Andes, lung capacity would develop in proportion to need. (Incidentally, such a change occurring in most people would add billions of liters of water per day to the atmosphere; the average person vaporizes about a liter per day.)

The thymus gland is extremely sensitive to stress, and atrophies quickly under the influence of cortisone or the sex steroids, but it can also regenerate under the influence of thyroid and progesterone (and under some circumstances, probably DHEA) or just good nutrition. Chronic atrophy of the thymus produces immunological weakness. Infectious diseases, then, are not so different from the degenerative diseases, since the failure of biological energy, and the resulting stress response, can be responsible for both classes of sickness. Even radiation damage, which seems like a fairly exotic problem in comparison to infections and aging, is ameliorated by things, like thyroxin, which increase energy production (Timchenko and Antipenko, "Conditions for the use of thyroxine as an antimutagen after total-body x-irradiation," *Radiobiologiya 21*(2), page 204, 1981).

Since the DNA repair process is energy dependent, greater biological energy prevents mutations. It used to be a common belief among biologists that the "wear and tear" of being alive would cause mutations, as well as other cell damage. However, many lines of evidence show that stress and energy depletion are common causes of chromosome damage.[5] Even free radical damage, which is popular as an explanation for aging, tends to become worse when energy is deficient[6] (fasting or emotional stress, e.g.), and not as a result of living itself.

Ideas of "original sin," "karma," "bad genes" or "wrong consciousness" are often invoked to "explain" suffering and disease. The most offensive example of this that I have heard was when a well known writer and lecturer claimed that people who are raped or murdered, even the victims of Auschwitz, including children, were responsible for their fate, through some mysterious spiritual defect. Even most of the conventional theories of aging and degenerative diseases contain a strong tendency to blame life itself as being at fault for its suffering.

Szent-Gyorgyi, in discussing some of his experiments with heart muscle, said "function builds structure, which increases the capacity for further function." The flow of energy through substance increases the order in that substance. More life and more energy can solve many of the basic problems of life.

REFERENCES

1. Storer, J. B. (1967). Relation of lifespan to brain weight, body weight, and metabolic rate among inbred mouse strains. *Exp. Geront., 2*(3), 173-182.

2. Jablonska, E. & Lamb, M. J. (1986). The inheritance of acquired epigenetic variation. *J. Theor. Biol. 13*, 69-83.

3. Rumyantsev, P. P. (1977). Interrelations of the proliferation and differentiation processes during cardiac myogenesis and regeneration. *Int. Rev. Cytol. 51*, 187.

   Oberpriller, J. 0., et al. (1988). Nuclear characteristics of cardiac myocytes following the proliferative response to mincing of the myocardium in the adult newt. *Cell & Tissue Research, 25*, 619.

   Claycomb, W. C. & Moses, R. L. (1988). Growth factors and TPA stimulate DNA synthesis and alter the morphology of cultured terminally differentiated adult rat cardiac muscle cells. *Dev. Biol., 127*, 257.

4. Polezhaev, L.V. (1972). *Loss and Restoration of Regenerative Capacity in Tissues and Organs of Animals.* Harvard Univ. Press.

5. Barbara McClintock (1984). The significance of responses of the genome to challenge. *Science, 226*, 792-801.

6. Meszaros, L. & Nemeth, A. (1982). Protection against lipid per-oxidation by substrate oxidation of mitochondria. *Acta Biochimica et Biophysica Academiae Scientiarum Hungaricae, 17*(1-2).

Protective effects of red light:

Gorbatenkova, E. A., et al. (1988). Photoreactivation mechanism of helium-neon laser on superoxide dismutase. *Dokl. Akad. Nauk. USSR,* 995-999.

Srivastava, A. K., et al. (1984). Induction of free radicals in seeds by high intensive flashes and effects on phosphorous metabolism. *Biol. Plant. (Prague) 26*(2), 88-98.

NOTES

* "The breeding rate is found to vary with the increasing scope and degree of care from parents. The decrease in the number of eggs and young is associated with a higher nervous organization expressing itself especially in intelligence, and helplessness is made concomitant with intelligence as its necessary condition." P. E. Davidson, p. 68 in *The Recapitulation Theory and Human Infancy*, Teachers College, Columbia University, 1914.

** "Vernalization" of grains might be thought of as suppression of neoteny, or activation of "restraint genes."

CHAPTER 18

THE TRADITION OF TRUTH

Several years ago I was thinking about the different levels of ability among famous painters of history, and I read about the family origins of some of them; the painters that I considered to have the best ability turned out to have come from poor families, in which skill in a craft was accepted as the way to earn a living. The bad--but famous--painters had “noble” origins, in which there was no tradition of skilled work. Painters who were “insiders,” familiar with the attitudes of the rich who paid for the pictures, knew what things would flatter, what things were fashionable, and what would contribute to prestige. The people from a lower class, however, had to be very skilled at painting if they were to succeed.

I think this illustrates something general about truth. A person’s vitality is drawn forward by meaningful work, that is, we grow to meet the demands of an important opportunity. Frans Hals’ paintings could only have been done by an alert mind, and one which had been alert for a long time learning the skills of drawing and painting. That skill revealed a large amount of truth. On the other hand, from Rubens we don’t learn much except that rich people ate well and dreamed of grandiosity.

William Morris said, “It is workmen only and not pedants who can produce real and vigorous art.” I think something similar should be said for science.

While some people explore and solve important problems, others do not explore, and even reject solutions to problems when they are made available by the workers who produced them.

One of my recurring objects of thought has been the slowness with which raw knowledge is assimilated. For example, I have been thinking about Broda Barnes's work on the prevention of heart disease with thyroid extract. He did solve much of "the riddle of heart attacks," but recent statements by the Heart Association show that the dominant forces in the health business haven't learned anything at all from his work, which he began 50 years ago. His work is clearly presented, not hard to understand, and it is scientifically so sound that no one challenges it, at least not on the scientific level. It is ignored, rejected by people who choose not to be bothered to read it. How many people have died from heart disease, since his work first became available? (And how many more from cancer, tuberculosis, and other diseases he showed occur mainly among hypothyroid people?)

I suggest that the resistance to Barnes's discoveries (and to nutritional therapies, etc.) has this explanation: that the health business is a Rubens-equivalent. There may be a family tradition of practicing medicine, but the child experiences family status long before he can experience the work of healing or the learning of special skills. Rather than an apprenticeship, it more closely resembles inherited nobility. Consider the character of most of the people who choose the profession, their motives, the way they are instructed, the meaning of truth in their lives.

But a few individuals within the profession live their lives for a different purpose, or cluster of purposes. They don't choose their problems in the way most likely to draw the favor of the powerful. Rather, there seems to be a need to understand, to take actions which lead toward the truth. I suspect that an "apprenticeship" has somehow been served in most of these cases.

Seen from a distance, it may seem that such people know in advance where the truth is, so they can proceed in the right direction. Deep involvement in the problem, investment of some of your life in the work, reveals that the truth is everywhere, and is clarified and arranged through the technical actions of the worker. The production of (or "discovery" of) truth is nothing mysterious, as most philosophers and some scientists would have us believe.

In the Spanish culture in which Picasso grew up, there was the category of "genius," and Picasso clothed himself with the attributes of genius; he fulfilled his father's expectations, believed that he was that, and found a culture willing to believe it.

At that point, he could do no wrong; everything he did was the work of genius-- "look at him, every scratch he makes with a pen is a stroke of genius." Criticism becomes impossible. The mystique around medicine is very similar. The public isn't qualified to criticize, and members of the profession are reluctant to criticize, because the myth must be preserved.

All the sciences, the military, and the activities of many politicians (the mystique of anti-communism in particular) propagate a way of living which just finds truth irrelevant. Mention "truth" in a roomful of scientists and you will hear a roomful of hoots and jeers, as if God had just walked to the podium at an atheists' convention.

But some people do take truth seriously, and a few try to find the "secrets" of people who have had it abundantly. Wilhelm Reich felt it had to do with a right relationship to the body and its energies; Einstein identified it with one's life. Robert Creegan, in *The Magic of Truth*, tries to clarify its nature, and to reveal where it is in our life. The physiologist Ukhtomskii was interested in the same problem, what it is that allows some people to directly see the important truth, while others are concerned only with self-assertion. Describing one kind of clear perception, Ukhtomskii says, "I learned that it is created by a great physical effort, a tradition handed down from others, and an attitude toward the world as toward a beloved, respected companion with whom one is close to the point of intimacy." He says "it is only via culture in the whole man, the culture in his deeds (will and love) that a vision can be achieved. . . ." Tradition, effort, attitude, and culture--I will be saying more about these later, in connection with meaningful work.

Creegan tries to say the same thing Ukhtomskii said 40 years earlier, some of his statements are worth quoting, just because he is a contemporary American. "The best way to assert that a step is possible is, more often than not, just to take that step." "It may be shown that all human effectiveness involves a special exhibition of truth." ". . . A plenitude of truth in action belongs to the fullest, most effective living." "The social reinforcement of unimaginative living, however, tends to restrict and retard the advancement of science itself." "Intellectual achievement is facilitated by exercises directing imagination away from obsessive channels." Rather than abstract exercises, Creegan might have considered the beneficial effects of meaningful work.

The Indo-European civilizations were formed by conquering many other, older civilizations about 3500 years ago; the survivors in the conquered cultures effectively became slaves, useful workers. Long after the Romans had conquered the Etruscans, they returned to the north and conquered newer European cultures. The productive cultures kept some of their roots, because that is where their usefulness came from-- their ability to do things well enough to sustain themselves and also the conquerors. The Romans busied themselves with military processes and recordkeeping.

All of the European civilizations had a similar internal division; at first different languages made the separation most obvious, but even after a kind of cultural unity had been imposed, the separation of productive and governing classes persisted. Sometimes the cycle was repeated several times (as in England), but always with the same overall outcome--if people were not highly productive, they were not worth conquering and enslaving.

Keeping this in mind, we can look at all kinds of cultural products, and see the results of this social division. In physics, we can compare the experimental attitude of Michael Faraday, the blacksmith's son, with the pompous errors of Lord Kelvin. In literature and the other arts, the richness of immediacy and realism in certain works distinguishes itself from the self-assertive fantasies and status obsessions of other works.

William Morris pointed out the similarity between the culture of the Romans and their subjects, and the culture of the workers and capitalists of his time, and he observed that these types expressed themselves in their art and science as different mental types. The productive people were described as constructive minds and "eye" people, or as "visionaries and practical people," whose work would be characterized by sensuous realism. The ruling classes he described as having analytical minds, concerned with formulas, labels, abstractions and theories, or formalisms.

Creegan and Ukhtomskii wrote about the two attitudes--full living vs. status concerns--and Morris observed that there are two cultures, sustaining those two attitudes, and showed the importance of recognizing their products all around us, so that we can learn to evaluate them, according to their origins and their purposes.

Our intelligence develops as we grow into the world, such as it is, and our world includes things that we learn about, and the people that we learn from. As we learn our language, we are learning the knowledge, the abilities, of the people who use that language. That is the culture that we learn, and it may contain elements that are true, useful techniques, and other elements which stunt us, negate our curiosity, or turn us back from the world.

Philippe Aries, in *Centuries of Childhood*, showed that "childhood" is a modern phenomenon, that children used to be thought of as little adults. From an early age they participated in the work of the family, and the life of the society. (Some related ideas are discussed by Jerome Bruner in *The New York Review of Books*, Oct. 27, 1983.)

If children are isolated from any intelligent work that the parents may perform, the world that they learn is one of functionless entertainment. Any tradition passed down by their parents is one of attitude, rather than of ability. If the child is allowed to be present when the parents work, the child wants to participate, to help, and to the extent that the children can participate, they are apprentices, and they learn effectiveness, personal power. If they are not apprentices, they still adopt the world of their parents as it is present in their speech, and they may receive a sense of power, but in this case their experience is that to *belong* is enough to be effective, that they "deserve" status. And if they don't receive this inherited sense of status, they may inherit helplessness.

If work is removed from the presence of children, how is the enabling truth which is in our culture to be distinguished from the useless aspects of the culture? If we lack a tradition of skilled work, then the point William Morris made becomes essential--we have to learn to reject the bad which is present in all the products of our culture--literature, painting, science, and so on.

To illustrate the double nature of our science, I will contrast two French scientists, de La Mettrie and Descartes.

Thinking of several kinds of biological facts, including the intrinsic motility or irritability of the intestine and heart, and the regeneration of the hydra from small fragments, Julien de la Mettrie proposed a new kind of physiology, based on the idea of organization. He suggested that thought was as natural for an organ with the structure of the brain, as beating was for the heart. He considered thought to be perfectly compatible with organized matter, and "to be apparently one of its properties--along with Electricity, Mobility, Impenetrability, Extension, and so on. . . . No: there is nothing base about Matter, except to those crude eyes which fail to recognize her in her most brilliant creations; and Nature is an Artist of unlimited capacities." He said that we should welcome our *natural* status, "cherishing life, scarcely understanding how feelings of disgust can taint the heart in this home of delight." (Compare Ukhtomskii's statement quoted above.)

One of de La Mettrie's physiological observations has been verified by recent experiments, nearly 250 years later-- he noticed that his thoughts were quicker when his pulse was accelerated by a fever.

According to Toulman and Goodfield (*The Architecture of Matter*), it was de La Mettrie's uncompromising materialism that caused his (truthful) physiology of organization to be forgotten by the scientific tradition, while Rene Descartes' double reality--mind and matter as separate substances, interacting only in humans, and there only at one point, which he thought might be the pineal gland--allowed him to compromise with religious beliefs. ". . . There had in fact to be a compromise; and the terms of this compromise have deeply influenced the subsequent development of science. Even today, their mark is evident in the structure of scientific ideas and institutions." (Page 165.)

For Descartes, science shouldn't mess with either God or the human mind. Theories about nature, including animals, could not refer to mentality. The traditional Christian sense of disgust for the body and matter had its influence on Descartes (and on the scientists who still follow his doctrines), but it was explicitly repudiated by de La Mettrie.

Above I have said that involvement in work orients a person toward the truth, that is, your "hypotheses" grow out of experience with the subject matter. Descartes explicitly claimed that it didn't matter how you choose your hypothesis: "Even supposing that I had assumed these principles at random, without having had reason to be convinced of their soundness, there would still be as good reason to suppose them the true causes of all that I have inferred, as in the case of a code deciphered by guesswork." A more perfect example of Morris's "theoretical" type would be hard to find. I don't know anything about de La Mettrie's childhood, but Descartes' father was a government official. I could give many contemporary examples, but these people were important in the history of philosophy and science, and they illustrate that both substance and method of science will vary systematically according to one's attitude toward the world.

If more scientists felt toward the world as La Mettrie and Ukhtomskii did, there would be not only a different "science," but a different world. "Nature is an artist of unlimited capacities," but such art is invisible to those who consider science to be a matter of decoding a message written once in dead matter.

THE FUTURE OF WORK

Several years ago, in the quarterly publication *Social Sciences*, I noticed an article by a man whose specialty was exploring the future of work; he projected a future in which a person's desire for growth and exploration is realized in his work. This person's job was to clarify the changes that must be made in the "economy" so that it will serve humanity--the workers and consumers--instead of vice versa.

Previously, in *Mind and Tissue*, I had briefly discussed some Soviet views on labor: That work tends toward perception, as machines become available; politics, work, culture, and science interpenetrate; brain function, education, science, and work have much in common--an emphasis on purpose and goals, deep reorganization, and complex perceptual interaction with the material. P. K. Anokhin and A. A. Ukhtomskii, and their students have created a sound basis for the role of goals and future thinking.

The attitude toward the future is an important part of how we orient ourselves and what concrete things we do to prepare for the future. A mechanistic view argues that we can't intervene to change the future, that it must fundamentally resemble the past, and that if people just invest in things that promise to give them a good profit the future will be nice. Another view sees the future as being composed of choices which lead to new choices, with new possibilities emerging as choices are put into action.

It's important that people start talking about the possible choices we have. If we accept that "the choice" is between being unemployed and having a job, the job we get is not likely to be what we want to do with our lives. And "status" isn't what I'm talking about. Giving maximum meaning to our lives should be one of the basic things that we demand of our work.

1. To start with concrete and familiar things, we might first want to discuss what work is, and why--under capitalism, and also under fascism, primitive cultures, and socialism. The issue of specialization could be considered here.

2. This might be followed by considering what work could become, and how. The nature of history, time, and culture should be considered, as well as the projections that are made by different groups.

3. And at some point, I think it is important to consider how work shapes us, how we are our work, and why it defines what we can be. Cultural, intellectual, and biological influences should be considered.

There are some things I want to quote, because they suggest some of the things that work is, what it does to us, and what it should be.

About 1790, William Blake wrote the poem "London," which begins:

> *"I wander thro' each charter'd street,*
> *Near where the charter'd Thames does flow,*
> *And mark in every face I meet*
> *Marks of weakness, markd of woe.*
> . . .
> *. . .the mind-forg'd manacles I hear."*

Another poem, "The Human Abstract," begins:

> *"Pity would be no more*
> *If we did not make somebody Poor. . . ."*

Repeatedly, Blake tried to define the mechanisms of oppression and limitation of the human personality. He observed that the State chartered corporations, licensed power; that it used false science, devious moralizing and religion, and illiteracy to create a culture of obedient drudgery. Commercial interests, he pointed out, distorted and degraded human life, art, and science:

> *"Schoolmaster of souls, great opposer of change, arise! O how couldst thou deform those beautiful proportions Of life & person; for as the person, so is his life proportion'd."*

> *"Thy self-destroying, beast form'd Science shall be thy eternal lot."*

Blake referred to factories as the "Satanic Mills," whose technology was invented:

> *"To perplex youth. . .& to bind to labours Of day & night. . .that they might file and polish. . .hour after hour, laborious workmanship, Kept ignorant of the use that they might spend the days of wisdom In sorrowfull drudgery to obtain a scanty pittance of bread. In ignorance to view a small portion & think that All. And call it demonstration, blind to all the simple rules of life."*

Several people in the following century were influenced by Blake's attitudes and perceptions, but most of them wanted to retreat to a simpler past, rather than (as Blake desired) to advance into a more generous future.

> *"And when all Tyranny was cut off from the face of the earth living flames winged with intellect and Reason, round the Earth they march in order, flame by flame. . . .Start forth the trembling millions into flames of mental fire. . . "Why sit I here & give up all my powers to indolence. . .?"*

That idea, that people would "ignite mentally" when oppression disappears, has become a recurring controversy. (J. D. Unwin vs. J. Prescott, for example.) In 1961, James Higgins was discussing the historical lethargy of people in Latin America, which is usually blamed on the tropical heat or an intrinsic character trait of the people. Describing the people he had met on previous trips to Cuba, who seemed lazy, careless, and sluggish, he described the change he saw after the revolution:

> *"And then they began to build, sing, dance, talk, to create dreams and start realizing them the next day, to plan, laugh, enjoy, to suddenly find that once they had knocked down certain artificialities, there were no longer any reasons whatsoever for not loving and helping one another. They broke through into the open and life became possible, that is, full of possibilities, real ones."*

Undoubtedly the invasion and blockade subsequently affected their energies, but for a time it was a remarkable phenomenon. J. P. Sartre described it as sleep being driven from the island along with the imperialists.

I think this "sense of possibility" is an important subject to study. Poets have led the way, but science can follow.

The "romantic" poet, Hugo von Hofmannsthal, said:

> *"The necessary is always possible. History proceeds in just that way, that something hardly credible is treated by a few as if it could be immediately realized."*

There is a sense that the future is open, or at least that it can be opened by our mental work. The musician Keith Jarrett said:

> *"People expect beautiful melodies. But I already know the melodies. So does everybody else. Rather than look for more beautiful melodies, everyone's purpose should be to find blind spots."*

Our present lives are usually divided between routine work and entertainment. The entertainment is supposed to enliven us, to help us recover from the deadening effects of routine work. Some people put great energy and concentration into their hobbies, because they find the activity intrinsically interesting. Such intrinsic value and interest is what should be demanded of our work. But for many people, free time is routinized too. To them, Jarrett's suggestion sounds like nothing but hard work. This is where the whole person has been affected by a certain approach to work, and work is seen as something to avoid--the idle rich seem to have found the only satisfactory life.

People like Blake, Higgins, and Marx have realized that there are different ways of being, that one is fragmented and diminished, and the other is whole, alive, and growing. When people feel that they are in possession of their own lives, then problems become opportunities. Each problem leads to new problems. The world draws us forward, and we are not defined by an "occupation" or "profession," but by the work we have achieved, and the problems we have confronted.

CHAPTER 19

THE EXPANDING EARTH

In 1959 I was assigned to teach a geography course that I wasn't prepared for, so I looked around the college's 100-year-old library for some ideas. In one old book, I saw an argument for an expanding earth, based on several types of geological evidence, including fossilized sand dunes which were supposedly too steep to have been formed under the present gravitational force. Later, when I looked at a map of the floor of the Atlantic Ocean, it seemed easy to interpret the roughly parallel ridges (which weren't known in the mid-19th century when the expanding-earth/increasing gravity idea had been proposed) as the traces of past volcanic insertions along a central crack, which had led to the separation of the Americas from Africa and Europe. Other people have continued to argue that the earth might have been much smaller in the distant past.

> *"Several geologists, notably Oakley Shields, have shown that not only the Atlantic margins but also those of the Pacific Ocean can be fitted together, with Australia against western South America, for example. But if the Pacific Ocean formed by the division of Australia and South America at the same time that the Atlantic formed by the division of Africa and South America, all continents must have been drifting away from each other. This, of course, is impossible unless the Earth, like the universe, has expanded within the last few hundred million years. Such expansion, in fact, has been ably argued by the Australian geologist S. Warren Carey."*[1]

Oddly, most steady-state universe theories which balance lost matter with newly created matter, tend to imagine the newly created matter as appearing only in stars or in empty space. I think there is something safely abstract and "academic" about theories which ignore the earth as a possible source of new stuff; any heat inside the earth apparently seems less threatening if it is merely the cooling embers of a cosmically remote fire.

When N. A. Kozyrev theorized that time's asymmetry might itself be a source of stellar energy, he predicted that planets would also have a steady source of internal heat in proportion to their mass, and his prediction matched the known heat of the earth, but it also predicted that Jupiter would be almost star-like in its heat emission, and that even the moon would produce some internal heat. He measured hot lunar emissions in 1960, and later space exploration confirmed several other major predictions of his. I think Kozyrev's work should at least make people recognize that even local matter is cosmic.[2]

V. I. Vernadsky, widely known as a crystallographer and biogeochemist, was able to see the cosmic, philosophical, and human implications of concrete geological things. Hannes Alfven, who argues that observable (i.e., in the laboratory) electrical forces can explain large processes, such as the development of form in galaxies, also confronts the conventional attitude which prefers to separate earthly science from cosmic science. Mendeleev, in the last century, and a few contemporary scientists, including the astronomer Thomas Gold, have believed that petroleum was formed deep in the earth by geological processes. These ideas, in giving a more creative role to the depths of the earth, have obvious cosmlogical significance.

Fred Hoyle and his colleagues have done some work which parallels that of Vernadsky, and which encourages a livelier way of thinking about cosmology. They have gathered spectroscopic evidence that cosmic dust (which makes up a significant part of the mass in the universe) more closely resembles bacteria than the other materials or particles which have been suggested (E. coli were used for spectroscopic comparison).

Rather than admit that certain facts are not explainable by any present conceptions, many so-called scientists will deny that such observations were made. Others will change the rules of reason, by subjectivizing science, resorting to operationalism, or to paradoxes such as wave-particle dualism.

Instead of leading to a science-mysticism, the existence of strange and conflicting observations should stimulate people to examine their assumptions, and to think new thoughts, and to devise new experiments.

In Frederick Soddy's last book, he suggested that cosmic rays might represent atoms which had been newly formed in interstellar space, and he mentioned that the permeation of space by ionized atoms, moving at high speed, would produce an evenly distributed microwave radiation. The existence of such radiation has been used by others to argue for the Big Bang story about the origin of the universe, so it is interesting that it was first mentioned in connection with a theory of continuous creation.

Early in his career, Soddy sided with Rutherford in deciding that the energy of radium came from within the atoms, and not from the interaction of the atoms with an energetic medium or ether. But the question of the new creation of matter, of interactions of particles in relatively empty space, has suggested to several physicists that the properties of "empty" space should be re-examined.

H. C. Dudley and others argued that a background composed of energetic neutrinos could explain many known processes as well as, or better than, present theories. Dudley argued, for example, that nuclear fission might be random only on a relatively gross scale, and that there might be microscopic domains which resonate in certain ways with the neutrino background. He argued that unsuspected simple conditions might be able to induce or "catalyze" the resonance, leading to fission or fusion under "mild" or ordinary conditions. Unknown to Dudley, an experimental physicist (Anderson) had demonstrated non-random nuclear decay in radioactive carbon when it was applied in a monomolecular layer to aluminum foil. (By chance, Anderson's and Dudley's papers--one experimental, the other theoretical--were published within a week of each other.)

Soddy's, Dudley's, and Kozyrev's intuitions about the ways in which new energy and matter appear in the universe try to combine nearly imperceptible phenomena (time, neutrinos, background radiation) with very important processes (stellar energy, nuclear energy, cosmic rays, the creation of matter). Creation is at the heart of existence, they might say, but is too often overlooked.

The idea of an expanding earth makes the intuition more immediate and concrete. S. W. Carey has surveyed the physical processes that might account for the geological facts which point toward expansion (indicating growth that is probably increasing exponentially, with a present annual increase of about 3 square kilometers of surface area), and he has proposed his own variant of physical theory to clarify his intuition.

But beyond the geological or cosmological issues, the expanding earth is a rich metaphor. The nature of star-energy, and the origin of life can't be such remote or abstract issues if we begin to sense the earth swelling, under and around us.

The nature of being is one boundary of the problem of understanding our own existence, and it is a moving boundary. "More, more!" is not (as Blake understood) a mistaken desire.

Another aspect of the problem of understanding has to do with our perception itself. Energized, intentional and personal, it is also responsive and public and endlessly flexible. It is part of the world, and the same motors that drive other processes must shape it. We might rephrase Le Chatelier's principle, in the way V. I. Vernadsky did, to say that systems use the energy that's available, and that our perceptions and social understandings are supported by energy flowing into the whole system. And again, there is no foreseeable limit to the intensity and richness of perception and understanding. The availability of appropriate energy--e.g., the right foods and the right fields and the right stimuli--would be the only limiting factor.

Though the "Dead Matter" theorists now dominate education, research, and medicine, and control so many other systems that the world is objectively in decline (toxins, radiation, deforestation, malnutrition, shrinking brains, extinction of species), techniques are now available to repair much of the damage and to start on an upward course. The repair and redirection begin within a system of passive constraints-- "a Mundane Shell."* The only thing that can resist the constructive organizing flow of energy is the absorptive, entropic principle of disorganization, and as Blake pointed out, there is a limit to contraction, but no limit to expansion.

S. W. Carey's image of the earth as an opening bud is a reasonable metaphor for our present situation:

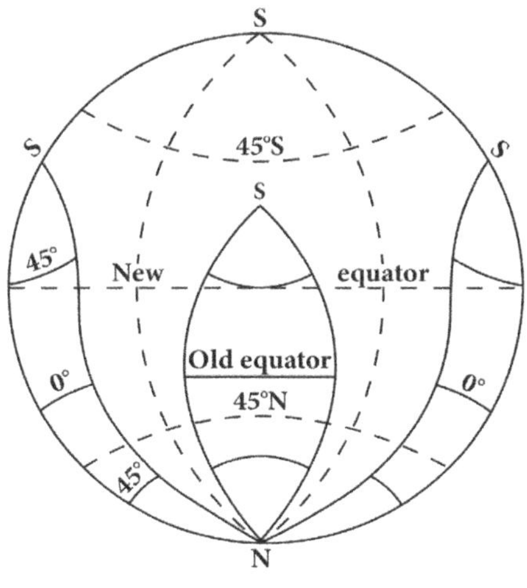

NOTE

*Blake's phrase "Mundane Shell" refers to the confining nature of fixed opinions and beliefs in our culture.

REFERENCES

1. Lowman, P. D., Jr. (1986). Faulting Continental Drift. *The Sciences.* 34-39.

2. Kozyrev, N. A. (1968). Possibility of Experimental Study of the Properties of Time. *JPRS: 45238.*

THE GATES OF PARADISE

At length for hatching ripe he breaks the shell.

One of Blake's images
of the Mundane Shell.

V. I. Vernadsky

ABOUT RAY PEAT

I have a Ph.D. in Biology from the University of Oregon, with specialization in physiology. The schools I have taught at include: the University of Oregon, Urbana College, Montana State University, National College of Naturopathic Medicine, Universidad Veracruzana, the Universidad Autonoma del Estado de Mexico, and Blake College. I also conduct private nutritional counseling.

I started my work with progesterone and related hormones in 1968. In papers in Physiological Chemistry and Physics (1971 and 1972) and in my dissertation (University of Oregon, 1972), I outlined my ideas regarding progesterone, and the hormones closely related to it, as protectors of the body's structure and energy against the harmful effects of estrogen, radiation, stress, and lack of oxygen.

The key idea was that energy and structure are interdependent, at every level.

Since then, I have been working on both practical and theoretical aspects of this view. I think only a new perspective on the nature of living matter will make it possible to properly take advantage of the multitude of practical and therapeutic effects of the various life-supporting substances--pregnenolone, progesterone, thyroid hormone, and coconut oil in particular.

"Marketing" of these as products, without understanding just what they do and why they do it, seems to be adding confusion, rather than understanding, as hundreds of people sell their misconceptions with their products. The very concept of "marketing" is at odds with the real nature of these materials, which has to do with the protection and expansion of our nature and potential. A distorted idea of human nature is sold when people are treated as "the market."

It seems that all of the problems of development and degeneration can be alleviated by the appropriate use of the energy-protective materials. When we realize that our human nature is problematic, we can begin to explore our best potentials.

----Ray Peat

Made in the USA
Monee, IL
13 September 2026

63066917R00090